VISUALIZING
BASEBALL

ASA-CRC Series on
STATISTICAL REASONING IN SCIENCE AND SOCIETY

SERIES EDITORS

Nicholas Fisher, University of Sydney, Australia

Nicholas Horton, Amherst College, MA, USA

Deborah Nolan, University of California, Berkeley, USA

Regina Nuzzo, Gallaudet University, Washington, DC, USA

David J Spiegelhalter, University of Cambridge, UK

PUBLISHED TITLE

Visualizing Baseball
Jim Albert

Errors, Blunders, and Lies: How to Tell the Difference
David S. Salsburg

VISUALIZING BASEBALL

JIM ALBERT

CRC Press
Taylor & Francis Group
Boca Raton London New York

CRC Press is an imprint of the
Taylor & Francis Group, an **informa** business

A CHAPMAN & HALL BOOK

CRC Press
Taylor & Francis Group
6000 Broken Sound Parkway NW, Suite 300
Boca Raton, FL 33487-2742

© 2018 by Taylor & Francis Group, LLC
CRC Press is an imprint of Taylor & Francis Group, an Informa business

No claim to original U.S. Government works

Printed on acid-free paper
Version Date: 20170720

International Standard Book Number-13: 978-1-1385-5115-2 (Hardback)
International Standard Book Number-13: 978-1-4987-8275-3 (Paperback)

Visit the Taylor & Francis Web site at
http://www.taylorandfrancis.com

and the CRC Press Web site at
http://www.crcpress.com

Contents

CHAPTER 1 ▪ History of Baseball   1

CHAPTER 2 ▪ Career Trajectories   17

CHAPTER 3 ▪ Runs Expectancy   33

CHAPTER 4 ▪ The Count   45

CHAPTER 5 ▪ PITCHf/x Data   55

CHAPTER 6 ▪ Batted Balls   67

CHAPTER 7 ▪ Plate Discipline   83

CHAPTER 8 ▪ Probability and Modeling   97

CHAPTER 9 ▪ Streakiness and Clutch Play   121

Bibliography   137

Index   139

Preface

Baseball and Data

Baseball is a special sport in its long-time relationship with data. From the beginning of American professional baseball in 1869, statistics such as runs scored, batting average, and counts of home runs were used to measure accomplishments of ballplayers. The baseball community now understands the importance of sabermetrics, the discovery, interpretation, and communication of patterns in baseball data, and all major teams have analytics groups as part of their payroll.

Although there is substantial interest in the study of baseball through statistical thinking, tables rather than graphs are primarily used by sabermetricians to communicate differences in players, season effects, and other types of effects. The author strongly believes that graphical displays, if properly drawn, are a wonderful way of communicating statistical patterns in baseball including distributions, relationships, and time-series patterns. Graphs are especially helpful in communicating to an audience who is not familiar with all of the detailed collection of observed and derived measures in baseball. These graphs can be used to enhance the statistical literacy of people who are interested in sports.

Availability of Data

Baseball is remarkable for its extensive collection of data, much of which is publicly available. Sean Lahman currently maintains season-to-season statistics for players, teams, managers, and post-season results and this data is available as a free download from baseball1.com. Retrosheet was founded in 1989 for the purpose of computerizing baseball game accounts. Currently box score data and play-by-play data is available for a large number of recent seasons from retrosheet.org. Much of the Retrosheet data is available on Baseball-Reference, one of the most comprehensive collection of baseball data. Data on pitches has been collected from the PitchFX system since the 2006 season and is easily accessible through the PitchRX package in R written by Carson Sievert. This system includes data on each pitch such as pitch speed and movement, location, and batting outcome. Most recently, the StatCast system, introduced in the 2015 season, collects data on player movements and batted balls including exit velocity, launch angle, and direction. This data is not generally publicly accessible, but summaries of StatCast data are available through web sites such as FanGraphs and Baseball Savant.

General Structure of the Book

This book is organized into nine chapters (inspired by the nine innings in a baseball game) and each chapter uses graphs to discuss a particular aspect of baseball history, illustrate basic principles of sabermetrics, and describe the new types of baseball data. Chapters 1 and 2 review the history of baseball, and the history of players as viewed by their career trajectories of performance. There has always been a tension between pitching and batting performance. For example, the rate of strikeouts is currently at an all-time high indicating that pitchers may be getting an upper hand in baseball. A graph of a player's career trajectory tells a story beyond the player's career statistics – one can see when the player achieved peak performance and highlight great seasons.

A baseball team wins a game by scoring more runs than its opponent, and many baseball performances can be measured by how many runs they contribute to the team. Chapter 3 uses graphs to introduce the concept of runs expectancy that can be used to measure the value of baseball plays. The notion of runs value is used in several chapters, to understand how to measure the batter or pitcher advantage in the count in Chapter 4, to estimate the probability of a team winning in the 5th inning up by two runs (Chapter 8), and to measure a batter's clutch performance in Chapter 9.

Chapters 5, 6, and 7 use graphs to illustrate the newer sources of baseball data. When one is watching a baseball, one typically sees a graph showing the pitch locations about the zone. Chapter 5 introduces the PitchFX data that provides pitch locations together with information on velocity, movement, and the pitch outcome. Chapter 6 introduces new batted ball data available through the StatCast system. This provides interesting insight into the direction, exit velocity, and launch angle of home runs, and one can estimate the probability that a batted ball is a hit based on knowledge of its exit velocity and launch angle. The FanGraphs site uses this new data to create a "plate discipline" collection of measures and Chapter 7 illustrates how swing and contact rates provide insight on strikeout and walk rates.

Outcomes of baseball games, baseball seasons, and individual season performances are uncertain and it is natural to use the language of probability to better understand the likelihoods of different outcomes. Chapter 8 uses a graph to explain how the probability of winning a game changes during the innings and how dramatic plays such as home runs can have a large impact on the probability of winning. Similarly, graphs are used in this chapter to explain the nature of baseball competition. A statistical model is used to describe baseball competition and simulations from the model allow one to relate team abilities with team performances such as winning the World Series. Graphs are also used to show how one can predict a player's final season batting average on the basis of his average after two months of the season.

Baseball fans are familiar with the current great players such as Mike Trout and Clayton Kershaw who have great season statistics. But more subtle batting and pitching performances are not as well understand, such as the tendency for a player to be streaky or consistently make clutch performances. Chapter 9 illustrates the use of graphs to show streaky hitting. It can be difficult to distinguish streaky and clutch performances from those performances from "random" data, and special graphs are helpful for detecting unusual streaky and clutch behavior.

Graphics Principles and Software

The graphs in the book were constructed following guidelines for visualizing data as expressed in Cleveland (2005). With respect to software, all of the graphs were produced using the ggplot2 package (Wickham, 2016) in the R statistical system (R Core Team, 2017). ggplot2 is a graphics system written by Hadley Wickham inspired by the grammar of graphics described in Wilkinson (1999). One of the goals of this book is to encourage readers to use ggplot2 graphics to illustrate baseball data and other types of data for their own enjoyment and research. Tutorial material on producing graphs for baseball data using R is given in Marchi and Albert (2013). R scripts for many of the graphs in this book and descriptions of how to obtain the relevant baseball data are available on the author's Github site `bayesball.github.io/VB` . Similar types of visual explorations of baseball data are available on the author's "Exploring Baseball with R" blog at `baseballwithr.wordpress.com`.

Intended Audience

This book was written for several types of readers. Many baseball fans should be interested in the topics of the chapters, especially those who are interested in the quantitative side of baseball. Many statistical ideas are illustrated in the book and so the graphs and accompanying insights may help to promote statistical literacy. From a practitioner's perspective, the chapters offer many illustrations of the use of a modern statistical graphics system, and the reader is encouraged to reproduce and hopefully improve the graphs in this book.

Acknowledgements

I am appreciative of my editor John Kimmel for his continuous support during the writing of this book. The comments of the reviewers were very helpful. Much of the writing was completed during a Faculty Improvement Leave at Bowling Green State University. Last, but certainly not least, I thank my wife Anne for her understanding and great patience during the book's completion.

Jim Albert, July 2017

CHAPTER 1

History of Baseball

INTRODUCTION

Professional baseball in the United States has a long history, as both the American and National Leagues have played since the 1901 season. Baseball from the beginning collected statistics, so most of the basic statistics such as counts of at-bats, hits, runs, home runs, strikeouts, and walks are available for all players since 1901. In this chapter graphs are used to explore patterns of rates of these basic statistics over time – these patterns help us learn about the history of baseball. The first part of this chapter focuses on team offensive statistics such as the number of runs or home runs hit by a team in a baseball game. Baseball is a competition between the offensive team and defense team and these plots show that the balance of this competition has drifted between the offense and the defense over baseball history. For example, we will learn that strikeouts per game are currently at an all-time high suggesting that pitchers currently have an advantage in modern baseball.

Since there is much interest in leaders in different statistical categories, the second part of the chapter focuses on the patterns of the leaders for different baseball measures. Graphs will be used to look at the leading batting averages, home runs, and on-base percentages. By focusing on the unusually high leading performances, one is introduced to some of the greatest players in baseball history.

TEAM STATISTICS

Runs Scored

A baseball team wins a game by scoring more runs than its opponent. One interesting aspect of baseball from a historical perspective is that the basic rules have not changed. A game is divided into nine innings, where each team has the opportunity to score runs. In a half-inning, players come to bat according to a prescribed batting lineup, and batters continue to come up until three outs are recorded.

Since runs are such a fundamental component of baseball, a good place to start is to explore the history of scoring runs. Figure 1.1 displays a time series plot of the average number of runs scored by a team in a game for each season from 1901 through 2015. The blue line is a smoothing curve that helps one see the basic patterns of growth and decline of run scoring.

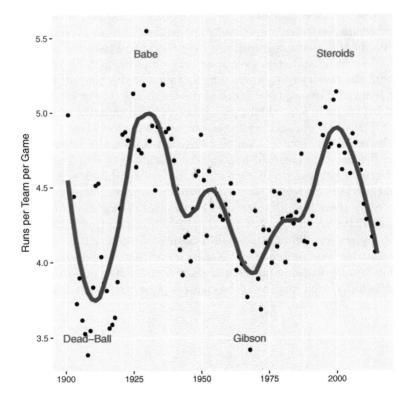

Figure 1.1 History of average runs scored in a game by a team.

Generally, one sees from the graph that teams tend to average between 3.5 and 5.5 runs a game, but there are dramatic changes in run scoring over the history of baseball. Labels are used to highlight four seasons 1908, 1930, 1968, and 2000 that were unusual with respect to run scoring.

- In the so-called Dead-Ball era (from about 1900 through 1920) in baseball, it was challenging to score runs. Games were held in spacious ballparks and the ball was "dead", partly by design and partly by overuse. The low season for scoring runs in this period was 1908. The batting average for all players that season was only 0.239 (contrasted to 0.255 in the 2016 season) and the Earned Run Average (ERA) for all pitchers was a low 2.37, contrasted with an ERA of 4.18 in the 2016 season. There were a number of dominant starting pitchers during this era including Addie Joss, Christy Mathewson, Cy Young, and Mordecai Brown.

- Run scoring dramatically increased after the Dead-Ball era, hitting a peak in 1930 when over 5.5 runs were scored on average by a team each game. The 1930 season is labeled in the figure with "Babe" as the dominant offensive player. Dominant hitters during this period included Babe Ruth, Lou Gehrig, Mel Ott, Al Simmons, and Chuck Klein. The 1927 New York Yankees, the "Murderer's Row", was arguably the most famous team during this period including Ruth and Gehrig. In contrast to the 1908 season, the average AVG in 1930 was a robust 0.296, and the average ERA was a high 4.81.

- After the Babe Ruth period, run scoring generally decreased from 1930 through 1968, although there was a modest increase in scoring from 1945 to 1955. The season 1968 was remarkable when only 3.4 runs were scored by a team in a game. This season was called the "year of the pitcher" and several pitchers had notable accomplishments. This season is labeled by "Gibson" since Bob Gibson had a remarkable low season ERA of 1.12 and Denny McClain won 31 regular season games. Carl Yastrzemski won the American League batting crown with a mere 0.301 batting average. After the 1968 season, the Rules Committee of Major League Baseball made several rule changes to allow for more offense. The strike zone was changed to the zone used before 1963 and the height of the pitching mound was lowered from 15 to 10 inches.

- Since the 1968 season, there was a steady increase in run scoring until the 2000 season. From the graph, we see the average runs scored by a team in the 2000 season was similar to the peak around the 1930 season. The 2000 season represented the period of time, the so-called "Steroids Era" when a number of players were believed to use performance-enhancing drugs. The top offensive players in the 2000 season were Todd Helton, Jason Giambi, Barry Bonds, and Alex Rodriguez.

- Since 2000, run scoring has dropped substantially, approaching 4 runs per team per game. It appears that baseball is again in an era similar to the Dead-Ball era that is dominated by pitching.

Home Runs

One of the dramatic ways of scoring runs is through home runs. Figure 1.2 displays a graph of the average number of home runs per team per game over the history of Major League Baseball. Although home runs directly produce runs, by comparing Figure 1.1 and 1.2, one sees that the historical pattern of home run hitting differs from the general historical pattern of scoring runs.

- Generally, there was a general growth in home run hitting over the period from 1901 through 2017.

- Looking closer, we see three time intervals when there was steady growth in home run hitting 1905–1940, 1945–1962, and 1975–2000. Also we notice two intervals when there was a decrease in home run hitting 1962–1975, and 2000–2015.

- The three periods of home run increase can be identified with great home run players. The growth of home runs in the 1920s and 1930s can be connected with the great hitter Babe Ruth. To illustrate the dominance of "The Great Bambino," during 1920 Ruth hit 54 out of a total of 630 home runs hit during that season.

  The growth in home run hitting between 1945 and 1962 peaked with great seasons of the home run hitters Mickey Mantle and Roger Maris. During the 1961 season, Mantle and Maris had a great dual to break Ruth's single-season record of 60 home runs. Maris did break Ruth's record with 61 home runs, a record that lasted for 36 seasons, but this record was believed not to be legitimate by the baseball commissioner since he played more games in the 1961 season than Ruth.

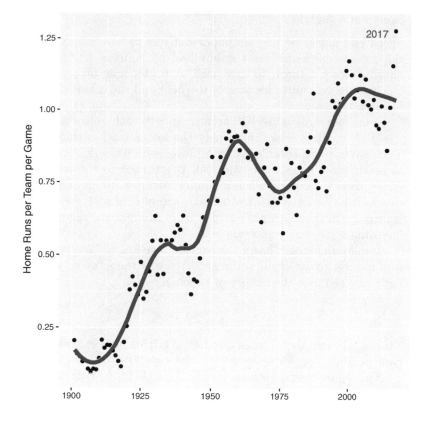

Figure 1.2   History of average number of home runs hit per team
per game.

The growth in home run hitting between 1975 and 2000
reached a climax with the "steroids sluggers" Mark McG-
wire, Barry Bonds, and Sammy Sosa. The 1998 season was
special in that McGwire and Sosa were competing to break
Maris' 37-year old home run record. Actually, both players
broke Maris' record in 1998 with McGwire hitting 70 home
runs followed by Sosa with 66.

At the time that this book was completed (summer 2017), we
observed an interesting surge in home run hitting. There was a 14%
increase in home run hitting between the 2015 and 2016 seasons
and currently (June 2017), teams are averaging 1.28 home runs
per game in the 2017 season which is the highest rate in baseball
history. It will be interesting to see if this pattern of increase in
home run hitting will continue in future seasons.

Speed in Baseball

Speed and quickness play an important role in baseball. A batter can get an infield hit on a groundball by running to first base quickly before a throw to first base. A fielder may need to react quickly to a ball hit to his area of the field and run a long distance to catch the ball.

Some measures of baseball performance directly relate to player speed. A triple is a base hit where the hitter reaches third base. Typically a triple is achieved by a player who has sufficient speed to reach three bases before the ball is returned to the infield. A stolen base is recorded when a runner advances to an extra base to which he is not entitled. Although the pitcher and the catcher contribute to the success of a stolen base attempt, the speed of the baserunner plays an important role.

Triples and stolen bases are exciting aspects of baseball, so it is of interest to explore how the occurrence of these "speed events" have changed over the history of baseball.

Triples

Figure 1.3 plots the average number of triples hit by a team in a game against year for seasons 1901 through 2015.

This figure clearly shows a steady decrease in the rate of triples over time. In the early years of baseball, teams would average 0.5 triples a game which means fans would see, on average, one triple a game. There was a dramatic decrease in the triple rate from 1925 to 1970, followed by a slight increase in the 1970s, followed by a gradual decline until the current season. Currently, the average number of triples per team per game is only about 0.17. Seeing a triple in a modern baseball game seems to be a rare event.

What are possible explanations for the decline in triples? In the Dead-Ball era, ballparks were more spacious than the modern ballparks, making it easier for a triple. Modern players do possess the speed to reach three bases quickly, but the dimensions of the modern ballpark may limit the triple opportunities. One contributing factor is defense – the modern outfielders may be more efficient in returning the ball to the infield. Also, a runner reaching second base is about as likely to score on a single as a runner on third base, so perhaps an attempt to reach third base for a triple is not considered a good strategy.

Stolen Bases

Another speed event in baseball is the stolen base. Figure 1.4 displays a historical view of the average number of stolen bases for a

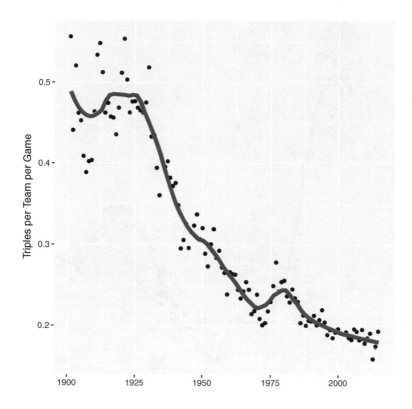

Figure 1.3 History of average number of triples hit per team per game.

team per game. During the Dead-Ball era, stolen bases were pretty frequent, averaging over a stolen base per team per game. As home runs became more prevalent, the rate of stolen bases plummeted, reaching a low value of 0.2 during the 1950 season. From 1950 until the early 1990s, the rate of stolen bases increased steadily to 0.75 per team per game. Since the early 1990s, the stolen base rates has shown a modest decrease and leveled off in recent seasons. It should be mentioned that there is much variation in stolen bases in modern baseball – in the 2016 season, Milwaukee had a total of 181 stolen bases compared with only 19 for Baltimore.

Why aren't stolen bases more common in baseball? First, being successful in stealing bases requires speed and training, and there are a relatively small number of players who frequently attempt to steal bases. Second, a stolen base attempt is a risky play and teams

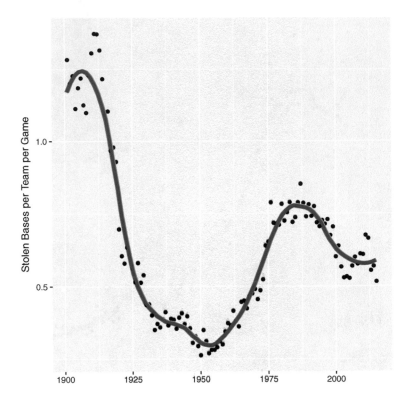

Figure 1.4 History of average number of stolen bases per team per game.

may decide not to attempt to steal due to this risk. Last, a team may decide on a strategy of scoring runs primarily on base hits, and stolen bases don't play an important role in this strategy. The current variation in team counts of stolen bases (remember the comment about Milwaukee and Baltimore) indicates that teams have different opinions on the value of stolen bases in producing runs.

The Other True Outcomes: Strikeouts and Walks

The "three true outcomes" in baseball are the three outcomes, home runs, strikeouts, and walks, that are completely determined by the pitcher-batter matchup and don't involve the defensive players on the field.

We have already looked at the historical pattern of home runs. What about the other two true outcomes? The top panel of Figure 1.5 graphs the average number of strikeouts per team per game across seasons.

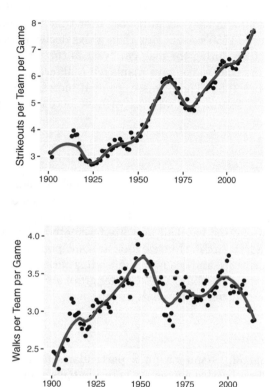

Figure 1.5 History of average number of strikeouts per team per game (top panel) and average number of walks per team per game (bottom panel).

It is clear from viewing this figure there has been dramatic changes in the strikeout rate over time. In the 1925 season, strikeouts were rare, but they steadily increased to 6 (a total of 12 strikeouts per game) in the middle 1960s. Then the strikeout rate dropped to about 5 in the middle 1970s. Since then, there has been a steady rise in strikeouts, approaching 8 (almost one strikeout in every half-inning) in the 2015 season. There was a remarkable strikeout record during the 2017 season when the Yankees and Cubs combined for a total of 48 strikeouts in a 18-inning game.

Since strikeouts often are related to walks (think about the great Hall-of-Fame pitcher Nolan Ryan who had a great fastball with control issues), one might anticipate a volatile pattern also in the history of walk rates. The bottom panel of Figure 1.5 displays the average number of walks per team per game across time.

Walks, like strikeouts, were rare in 1900 (only about 2 walks per team per game) and they steadily rose over time, reaching a peak of 3.5 to 4 in the 1950 season. But then walks declined and reached a low value of 2.8 during the magical "year of the pitcher" in 1968. It is interesting that walks have displayed a different pattern than strikeouts since 1968. There was a gradual increase in walk rates until the 2000 season but walk rates have actually decreased in recent seasons.

LEADERS

The first set of graphs give one a general understanding how run scoring, home runs, and speed statistics have changed over the history of baseball. But these graphs don't highlight accomplishments of individual players; baseball fans are fascinated with the players listed on "leaderboards". Here we look at some popular measures of hitting performance and see how the leading measure has changed over the history of baseball. These leader graphs will highlight some of the greatest hitters in baseball history.

Batting Average

The MLB batting champion for a particular season is the player with the highest batting average, where a batting average is defined by the number of hits divided by the number of at-bats:

$$AVG = \frac{H}{AB}$$

Figure 1.6 plots the AVG of the batting champion for all seasons from 1901 through 2015. Special high batting averages are labeled with the player's last name. Looking at the smoothing curve, we see some interesting patterns:

- The leading batting average showed a general increase until the 1925 season.

- In the next 43 seasons between 1925 and 1968, there was a steady decrease in the leading batting average.

- From the mid-1960s until the 2000 season, there was a gradual increase in the leading average, and the average has displayed a steady decrease until the 2015 season.

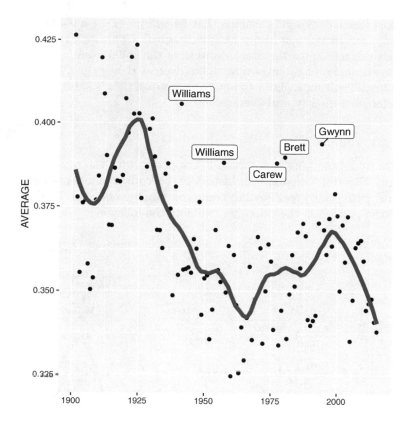

Figure 1.6 History of the leading batting average. Special batting averages are labeled with the last name of the player.

After thinking about the general pattern, one looks for interesting deviations from the pattern. In the early years of baseball (seasons 1901 through 1925), a batting average of 0.400 or higher was relatively common. Looking at Figure 1.6, one sees 10 leading averages over 0.400. The most recent average at this level, labeled by "Williams" in Figure 1.6, was Ted Williams' 0.406 average in the 1941 season. The figure also labels the four leading batting averages exceeding 0.350 since the 1950 season:

- Ted Williams 0.388 average in 1941

- Rod Carew's 0.388 average in 1977

- George Brett's 0.390 average in 1980

- Tony Gwynn's 0.394 average in 1994

Most baseball experts conclude that we will never see a 0.400 batting average again in Major League Baseball. Gould (2011) argues that the reason for the disappearance of the 0.400 average is that the variability of batting averages has decreased over time, and it is more difficult for a player to obtain an average that is substantially different from a typical average.

Home Runs

The home run king in a season is the player with the most home runs. Figure 1.7 displays the history of the leading number of home runs with some great leading home run numbers labeled. Looking at the pattern of the smoother, we note the following:

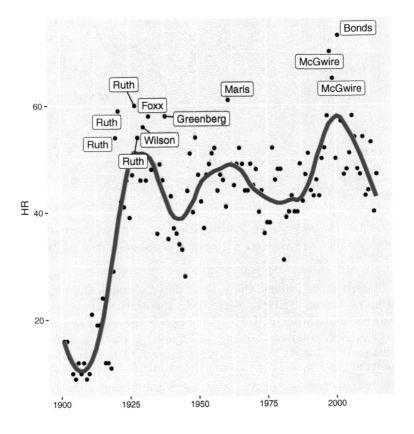

Figure 1.7 History of the leading number of home runs. Special home run counts are labeled with the last name of the player.

- A small number of home runs were hit in the so-called "Dead-Ball" erbetween 1900 and 1920.

- Since 1920, the leading number of home runs has been relatively constant over time.

- There are three general peaks in the leading home runs, around 1925, 1960, and 2000, and valleys around 1945 and 1985.

There are a number of leading home run accomplishments that are labeled in Figure 1.7. Babe Ruth, arguably the greatest hitter in baseball history, hit 54, 59, 60, and 54 home runs in the 1920, 1921, 1927, and 1928 seasons. Other players in Ruth's period had similar leading home run counts: Hank Wilson's 56 home runs in 1930, Jimmie Foxx's 58 home runs in 1932, and Hank Greenberg's 58 home runs in 1938.

After the great home run leaders in the 1920s, the next great home run leader was Roger Maris who set a record with 60 home runs in the 1961 season. After Maris, the leading home run count stayed relatively constant until the so-called "steroids era" about the year 2000. The famous leaders in this period were Mark McGwire who led MLB with 70 and 65 home runs in the 1998 and 1999 seasons, and Barry Bonds who set the current record of 73 home runs in the 2001 season.

On-Base Percentage

A batting average is one indicator of a player's success in hitting. But a batting average, by definition, ignores outcomes such as walks and hit-by-pitches that allow a batter to reach first base. A better measure of a player's ability to get on base is the on-base percentage

$$OBP = \frac{H + BB + HBP}{AB + BB + HBP + SF}.$$

This particular hitting statistic was made popular in the book *Moneyball* by Lewis (2004) that was later made into a movie. During the 2002 season, the OBP was an underappreciated measure and the Oakland general manager could cheaply sign players who had the talent to get on base. Figure 1.8 gives a historical view of the leading on-base percentage. One sees that the leading OBP was small during the Dead-Ball era and rose until the 1925 season. Then the OBP value dipped a bit and had a dramatic drop during the 1960s. From 1968 until 2000, there was a steady increase in the leading OBP and the value has dropped steadily in recent seasons.

Figure 1.8 labels the leading OBP values exceeding 0.500. There

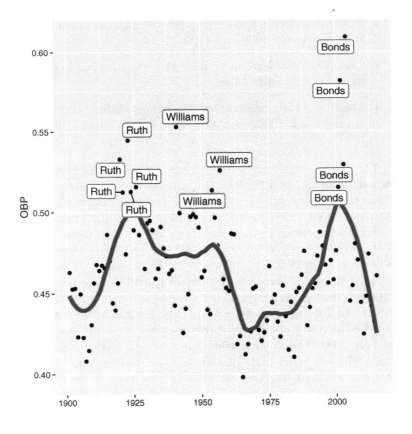

Figure 1.8 History of the leading on-base percentage. Special on-base percentages are labeled with the last name of the player.

were 12 player/seasons labeled and remarkably they were achieved by only three players.

- Babe Ruth, the great home run slugger, had OBP seasons of 0.532, 0.512, 0.544, 0.513, 0.516 in the 1920, 1921, 1923, 1924, and 1926 seasons.

- The next OBP titan was Ted Williams who had OBP values of 0.553, 0.513, 0.526 in the 1941, 1954, and 1957 seasons.

- The final player Barry Bonds had OBPs of 0.515, 0.582, 0.529, 0.609 in the 2001, 2002, 2003, and 2004 seasons. In the 2004 season, Bonds had a record 232 walks in 617 plate appearances for a remarkable 0.609 on-base percentage.

Although many current players are very good in getting on-base, it is doubtful that one will see OBP stars at the level of Ruth, Williams, and Bonds in the forseable future.

WRAP-UP

This chapter has demonstrated that although baseball has been playing by the same basic rules for over 100 years, the patterns of different types of batting events has shown big changes over baseball history. Generally, home run hitting and strikeouts have increased over time, while the rates of other events such as triples have decreased substantially from the 1900 season to the current day. Also our historical exploration of leaders for specific measures has helped in identifying some of the greatest hitters in baseball history.

One implication of these changes is that one cannot judge the greatness of a particular batting measure just by looking at its value. For example, due to the changes in the average batting average over time, it is difficult to evaluate a player's hitting performance by simply looking at his batting average. A leading batting average of 0.400 would be remarkable in modern baseball, but 0.400 hitters were relatively common in the 1920s. So any player's performance should be evaluated in the context of the seasons when he played. This advice is especially important when players from different eras are under consideration for admission to the baseball Hall of Fame. When players are compared, one needs to adjust any career hitting measures for the era in which they played. Otherwise a player might get into the Hall of Fame not because they were great but because they happened to play during an era when pitching or batting were dominant.

CHAPTER 2

Career Trajectories

INTRODUCTION

Chapter 1 illustrated the changes of baseball team performance over the history of baseball, but interesting patterns in performance exist over time when viewed at the individual level. Generally a professional athlete enters the professional ranks at a modest level, increases his or her level of performance until a peak level is reached, and then decreases in performance until retirement. When some measure of performance is graphed against age, this graphical display is called the *career trajectory* of the athlete. Although the general "rise, hit a peak, and then fall" career trajectory pattern happens in all sports, the characteristics of this pattern can vary. For example, some athletes may peak at a younger or older age, and the rise and fall from peak performance can occur at different rates.

In this chapter, career trajectory patterns in baseball are explored. By looking at some representative graphs of some famous players, one gains a general understanding of trajectory patterns in baseball. By viewing these graphs, one learns much more about the season-to-season accomplishments of a player than the cumulative statistics typically used to summarize a player's career.

UNDERSTANDING CAREER PERFORMANCE

How Do Baseball Players Age?

Suppose one is interested in learning about how baseball players age. Here are some basic observations about player ages and the statistics one collects.

- Practically all of the ages of professional baseball players fall between 20 and 40 years. The better players tend to make their Major League Baseball debuts close to 20 and are more likely to play until age 40 or later.

- One typically keeps track of career statistics. For example, Barry Bonds had 762 career home runs, Ty Cobb had a career batting average of 0.366, and Cy Young had 511 career wins.

We are interested in how a player's performance is spread out over all the seasons of his career. If one plots some measure of performance against a player's age, the resulting graph is called the **career trajectory** of the player.

Barry Bonds Home Run Trajectory

Barry Bonds is the current career leader in home runs with 762. He had a long career in professional baseball from age 21 to 42. Looking at Bonds' 762 home runs over his 22-season career, one can compute that he averaged 34.6 home runs each season.

If Bonds' home run hitting performance was constant over his 22 seasons, then ignoring the typical season-to-season variation, the summary pattern of home runs would look like Figure 2.1.

But if one actually plots Bonds' home run counts against season in Figure 2.2, one sees that Bonds' pattern of hitting home runs has a distinctly non-constant pattern.

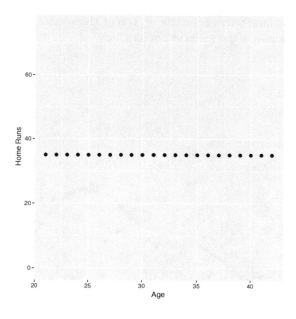

Figure 2.1 Barry Bonds' career trajectory if his home run hitting performance was constant over seasons.

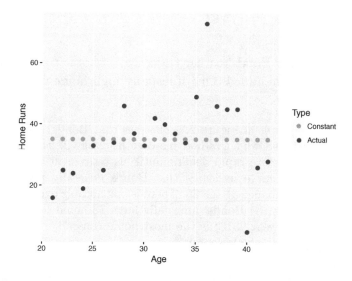

Figure 2.2 Barry Bonds' actual career trajectory of hitting home runs compared with a constant pattern of performance.

A smoothing curve is drawn on top of the scatterplot in Figure 2.3 to help us see the general pattern.

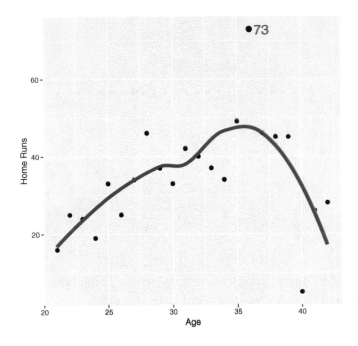

Figure 2.3 Barry Bonds' career trajectory of home runs with a smoothing curve added. One unusually high home run count is labeled.

From the graph in Figure 2.3, one sees that Bonds' home run count generally shows a gradual increase from age 21 through 36 and then has a more rapid decline until his retirement age of 42. The smoothing curve indicates that Bonds generally peaked in home run performance at age 36. There is one notable exception to the general pattern – Bonds' unusually large 73 home runs at age 36 currently (in year 2016) is the most home runs hit by a Major League player in a single season.

COMPARING CAREER TRAJECTORIES OF HITTERS

It is interesting to explore the career trajectory of a single player such as Barry Bonds. But to gain a general understanding of aging patterns in baseball, one wants to compare career trajectories of "similar" players. We discuss the choice of performance measure and how one determines if two players have similar careers.

First, one needs to decide on the measure of performance. There are many possible choices for a measure. For example, if one was interested in slugging performance, one could use the count of home runs, the rate of home runs (home runs divided by plate appearances), the slugging percentage, or other measures. If one was interested in general hitting performance, a multitude of batting measures are possible. Currently there are particular measures of performance that provide good summaries of offensive performance. We focus on the use of $oWAR$, the offensive wins above replacement. Essentially, this measure indicates how many wins the player contributed to the offense of his team over what would be contributed by a suitable replacement player. A basic career trajectory graph plots the season $oWAR$ performance measure against the player's age.

How does one find a group of similar players? Bill James (1994) introduced the concept of similarity scores to compare two baseball players. To compare two players, one starts at 1000 points and then makes subtractions based on the differences in the career statistics in different categories such as games played, runs scored, hits, and so on. There is an adjustment for the fielding position in the calculation. Generally it is desirable to compare players who played the same or similar fielding position.

What patterns do we expect to see in a career trajectory? One expects a player to mature and increase in performance during a particular year and then decline until retirement. Conventional wisdom says that baseball players peak between 28 and 32, although Bill James in his *Baseball Abstracts* writing gave evidence to indicate that the range was more like 25 to 29. Personally, I believe that there may be a general pattern of aging, but players age differently depending on their playing position, physical condition, and other patterns. But it is helpful to think of 30 years as a benchmark when one explores these career trajectory graphs.

In the comparisons to follow, a famous baseball player will be chosen and James' similarity scores are used to find three other players who are similar with respect to career performance and fielding position. The comparisons of career trajectories of $oWAR$ are helpful in understanding general patterns of aging in offensive performance.

Mickey Mantle and Three Similar Hitters

Mickey Mantle was one of the great sluggers while I was growing up. He had a great combination of speed and power and he was one of the leaders of the New York Yankees when they were winning a number of World Series in the 1950s and 1960s. Using similarity scores, three players with similar batting statistics to Mantle were Frank Thomas (great player for the White Sox from 1990 to 2008), Eddie Matthews (third baseman for the Braves between 1952 and 1968), and Mike Schmidt (great Phillies third baseman who played between 1972 and 1989).

Figure 2.4 displays career trajectories of Mantle, Thomas, Matthews, and Schmidt using the $oWAR$ statistic. One sees from

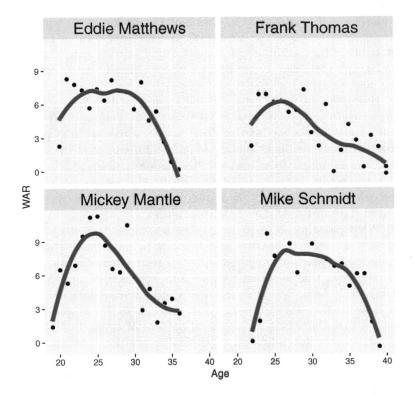

Figure 2.4 oWAR trajectories of Mickey Mantle and three similar players.

this figure that Mantle matured rapidly and appears to peak about age 25 and then generally declined in performance until his retirement at age 36. Frank Thomas had a similar trajectory, peaking about age 26, but his performance at his peak is noticeably smaller than Mantle's. Both Matthews and Schmidt had longer periods of best performance. Matthews displayed a constant level of offensive performance from age 21 to 30 and then declined until retirement. Similarly, Schmidt's performance was very consistent from age 24 through 37. Comparing the four players with regard to their peak, we see that Mantle had the best peak performance with an $oWAR$ statistic value close to 12 at age 25.

Babe Ruth and Three Similar Hitters

Next consider Babe Ruth, the "Sultan of Swat" who played for 22 seasons, 15 of them for the New York Yankees. Figure 2.5 displays the career trajectory of Ruth and three similar players, Barry Bonds, Ted Williams, and Lou Gehrig.

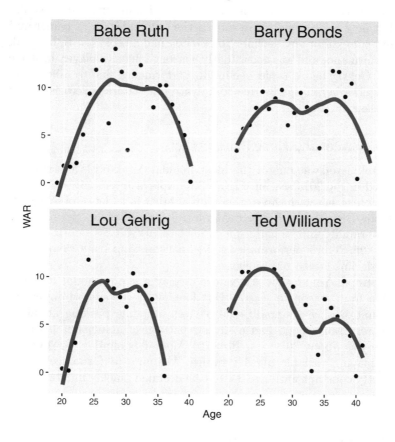

Figure 2.5 oWAR trajectories of Babe Ruth and three similar players.

Ruth had a remarkable career – one sees that his $oWAR$ value remained about 10 runs for the period 25 to 35 years. Since Ruth actually started professional baseball as a pitcher, his $oWAR$ values were small for the first few seasons of his career, and Ruth declined rapidly between 35 and 40. We have earlier commented on Barry Bonds' unusual career trajectory for home run counts. If one removed the three large $oWAR$ values, then Bonds' trajectory would be typical with a peak about 30 years. But Bonds had those three high $oWAR$ values at ages 36, 37, and 39 – this is very unusual for a baseball slugger.

Ted Williams had a career interrupted by military service for the ages 24 through 26 – from a baseball perspective, this was unfortunate since his best $oWAR$ seasons occurred early in his career. His offensive performance declined somewhat after age 30, but he was a strong hitter during the remainder of his career and did not exhibit any decline. Lou Gehrig had a rapid rise to his peak performance and was consistently a strong offensive player until age 34. Gehrig had a rapid decline in performance due to illness and passed away at age 37 due to amyotrophic lateral sclerosis (ALS) disease.

Derek Jeter and Three Similar Hitters

Derek Jeter was one of the most popular Yankee players who retired in the 2014 season. Figure 2.6 displays Jeter's $oWAR$ trajectory together with three other similar great infielders Robin Yount, Craig Biggio, and Roberto Alomar. Jeter's offensive performance rose rapidly and he had some of his best $oWAR$ seasons at ages 24 and 25. Jeter's performance stayed constant until age 35 and then he declined until retirement.

Robin Yount's $oWAR$ values display a good amount of variation from season to season. But looking at the smoothing curve, Yount displayed a traditional pattern of aging, peaking about 27 years. Craig Biggio had a similar pattern of aging, but his peak age was about 30 years. Roberto Alomar is similar to Yount in displaying variable $oWAR$ values. Alomar's performance stayed pretty constant until age 34, but he declined quickly in later years.

General Patterns of Aging

We have seen different patterns of aging in baseball batters and have commented on the player's perceived peak age, the age at which the player achieves peak performance. Can one make general comments about peak ages? All of the players were identified who had at least 5000 career plate appearances. Each player's ca-

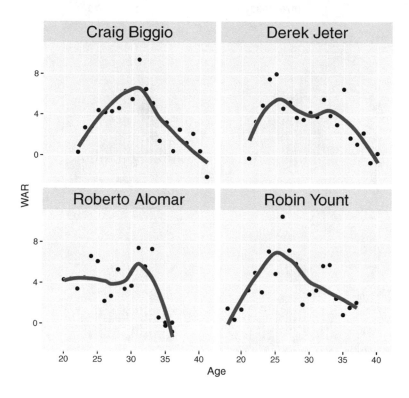

Figure 2.6   oWAR trajectories of Derek Jeter and three similar players.

reer trajectory of batting was graphed where the measure of performance was $wOBA$, the weighted on-base percentage, a similar measure of overall batting performance to $oWAR$. For each graph, the age of peak performance was found by identifying the age where the smoothed curve had its largest value. For example, looking at Figure 2.4, Mickey Mantle's peak age using this method would be 25. Figure 2.7 displays the distribution of the peak ages for the 886 players, where the baseball era has been divided into one of the four groups 1870–1930, 1930–1970, 1970–1990, and 1990–2010) Although the peak age varies across players, the typical peak age is about 28 and about 30% of the players peak between 27 to 29 years. It should be noted that these conclusions are based on the players who have had long careers and may not reflect the peak ages for players with short careers.

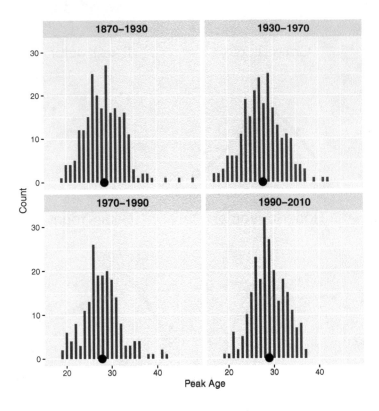

Figure 2.7 Peak age distributions for MLB players with at least 5000 career plate appearances, divided by the era in which they played. The mean peak age for each group is displayed by a solid dot.

COMPARING CAREER TRAJECTORIES OF PITCHERS

Career trajectory graphs are also helpful for understanding the career performances of pitchers. Measuring pitcher quality is more difficult than measuring batter quality since the pitcher works together with his defense in preventing runs. One is unsure if a pitcher allows a hit due to a bad pitch or a bad defensive play. With this caution, we use a popular and easy-to-understand pitching measure, the WHIP which is defined as the number of hits and walks allowed per inning:

$$WHIP = \frac{H + BB}{IP}.$$

Here a strong pitching performance is equivalent to a small value of WHIP. So "peak" performance corresponds to the smallest value of WHIP during a pitcher's career.

Four Famous Lefties

We begin by comparing four famous left-handed starting pitchers. Johan Santana pitched for 12 seasons between 2000 and 2012 for the Minnesota Twins and the New York Mets, winning the American League (AL) Cy Young award in 2004 and 2006. Ron Guidry played for the New York Yankees for 14 seasons between 1975 and 1988, winning the AL Cy Young award in 1978. Sandy Koufax pitched for the Los Angeles Dodgers for 12 seasons from 1955 through 1966, winning the AL Cy Young award three times. Steve Carlton had a long career of 24 seasons, 14 with the Philadelphia Phillies, winning the Cy Young award four times. Figure 2.8 displays career trajectories of WHIP against age for these famous lefties.

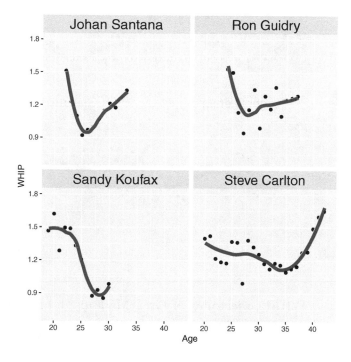

Figure 2.8 WHIP trajectories of Johan Santana, Ron Guidry, Sandy Koufax, and Steve Carlton

Although all four pitchers had great careers, the patterns of their trajectories are different. Both Santana and Koufax started slow but had great WHIP performances for a few seasons. Santana's WHIP steadily increased until retirement; in contrast, due to a shoulder injury, Koufax retired from baseball at his peak level at age 30. Guidry's WHIP values show a lot of variability, but it seems that his best performance was early in his career. Carlton had a long career and actually seemed to peak around age 35.

Three Braves and the Rocket

Figure 2.9 compares the WHIP trajectories of four great starters, Greg Maddux, John Smoltz, Roger Clemens, and Tom Glavine.

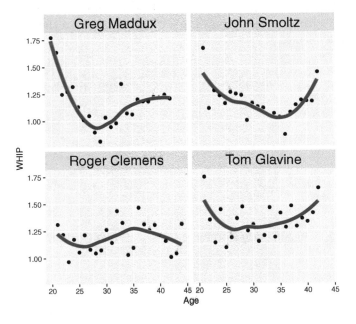

Figure 2.9 WHIP trajectories of Greg Maddux, John Smoltz, Roger Clemens, and Tom Glavine

Maddux, Smoltz, and Glavine all played for the Atlanta Braves in the 1990s. Maddux had four Cy Young awards in his 23-year career from 1986 to 2008. Smoltz won one Cy Young award in a long 22-year career, and Glavine won two Cy Youngs in a 22-year career. Comparing Maddux and Smoltz, Maddux appeared to peak (with respect to WHIP) about age 28; in contrast, Smoltz peaked more about age 35. Glavine had a pretty consistent pattern of WHIP values although the general level of values seems lower than Maddux and Smoltz.

Roger Clemens (the "Rocket"), winner of seven Cy Young awards in a distinguished 24-year career with the Boston Red Sox and other teams, has an interesting trajectory. His WHIP values reached a minimum near age 26 and his performance seemed to get worse until age 25. But he concluded with some of his best WHIP seasons toward the end of his career.

Four More Great Righties

Figure 2.10 compares WHIP trajectories of Dwight Gooden, Pedro Martinez, Roy Halladay, and David Cone.

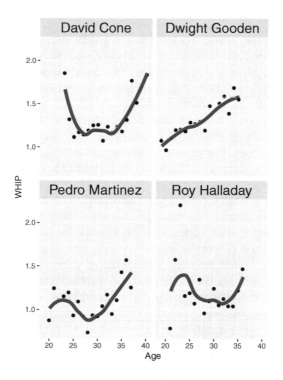

Figure 2.10 WHIP trajectories of David Cone, Dwight Gooden, Pedro Martinez, and Roy Halladay.

Gooden pitched for 16 seasons, primarily for the New York Mets, between 1984 and 2000. He won a single Cy Young award at age 20. Martinez pitched for five teams for 18 seasons between 1992 and 2009. In the span of four seasons, he won three Cy Young awards. Halladay pitched for the Toronto Blue Jays and the Philadelphia Phillies for 16 seasons from 1988 through 2013 and won two Cy Youngs. Cone played for five teams in a 17-year career and was a member of five World Championship teams including four with the New York Yankees.

Comparing the trajectories, Goodon clearly was best as a young pitcher and gradually got worse in WHIP over his career. Martinez had a more typical trajectory shape, peaking in his late 20s. Halladay's WHIP values were less consistent, but some of his best seasons were between 31 and 34. Cone was a consistently strong pitcher from ages 25 through 35.

WRAP-UP

In this chapter, career trajectories of performance have been visually explored for several great hitters and pitchers from baseball history. Most of these players had long careers that make it easier to examine their season-to-season performances from their rookie seasons to retirement. Although players typically achieve peak performance between age 28 and 32 years, this chapter has demonstrated that peak performances can vary between hitters and between pitchers. For example, Mickey Mantle and Robin Yount clearly peaked in their mid 20s while Babe Ruth and Barry Bonds exhibited high levels of performance in their mid 30s. Also the rate of increase until peak level and the rate of decrease until retirement vary among players. For example, the rate of decrease in WHIP to retirement appears to be greater for David Cone than Dwight Gooden (Figure 2.9).

Do these career trajectory patterns of players matter to general managers in MLB? Every winter teams have to make decisions about offering contracts to free agents and these decisions should take into account the age of the free agent and the predicted pattern of future performance. In the winter of 2011, the Angels signed Albert Pujols to a 10-year contract for $240 million, and then Pujols exhibited a big drop in the level of batting performance in his years with the Angels. It is difficult to predict a player's future performance, but Pujols was signed at age 31, and based on general knowledge about career trajectory patterns, one should have anticipated some drop in batting performance after age 31. Of course, decisions about contracts depend on many factors, but knowledge about the patterns of trajectories of similar players should impact these decisions.

CHAPTER 3

Runs Expectancy

INTRODUCTION

A baseball game is won when one team scores more runs than the opposing team. The ratio of a team's total runs and total runs allowed has a strong relationship to the ratio of wins to losses. Since runs are such an important aspect of a team's performance, it is helpful to think of the value of plate appearance events in terms of runs. Runs expectancy is a measure of the opportunity to score during an inning given a particular number of outs and runners on base.

Graphs help us understand the pattern of run scoring, and the pattern of expected runs for different numbers of outs and runners on base. The overall value of baseball plays, such as a double or an out, can be estimated using runs expectancy. Going further, one can show how runs expectancy can be used to determine how the value of a hit, such as a home run, depends on the game situation. Moreover, since we can estimate change in the runs expectancy after any play, one can base evaluations of player performances for both batters and pitchers using runs expectancy.

WHAT IS RUNS EXPECTANCY?

Runs Scored in a Half-Inning

A baseball team comes to bat in a half-inning – how many runs will they score? During the 2015 season, there were 45,315 half-innings and Figure 3.1 displays a bar chart of the number of runs scored in all of these half-innings. Most of the time, about 73% to be precise, the team won't score any runs, and the team scores exactly one run about 15% of the time. It is pretty rare (about 12% of the time) that a team will score more than one run in a half-inning.

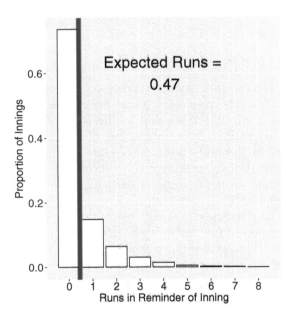

Figure 3.1 Bar graph of the number of runs scored in an inning for the 2015 season. The expected runs scored is displayed by a vertical line.

Suppose one finds the number of runs a team scores, on average, in a half-inning by finding the total number of runs scored and dividing this sum by the number of half-innings. One obtains the average of 0.47 runs. We will refer to 0.47 as the expected runs scored starting at the beginning of an inning with 0 outs and no runners on base. This expected runs scored is shown by a vertical line on the graph in Figure 3.1. Of course, a team can't score 0.47 runs in an inning. But if a team plays many innings, then on average, they will score about half a run per inning.

Advantage of a Runner on Second Base

A team's potential to score runs will change depending on the game situation. Suppose the first batter in an inning gets a double, so now there is a runner on second base with no outs. Given this new situation, the team is more likely to score runs. During the 2015 season, there were 3281 instances where a team had a runner on second base with no outs. Of these situations, we keep track of the number of runs scored in the remainder of the inning. A bar chart of the number of runs scored starting from the runner on second and no outs state is displayed in Figure 3.2. Now we see that there is about a 38% chance of scoring 0 runs, a 36% percent chance of scoring one run, and there is an over 26% chance of scoring more than one run. Just like the first situation, one can summarize this distribution by computing an average runs scored of 1.11, displayed as a vertical line in the graph. This number represents the expected number of runs scored in the remainder of the inning starting from a runner on second and no outs state.

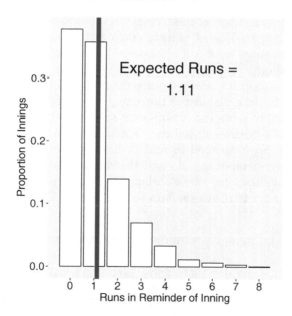

Figure 3.2 Bar chart of the number of runs scored in the remainder of the inning when there is a runner on second base with no outs. The expected runs scored is displayed by a vertical line.

Value of a Double: Change in Runs Expectancy

At the beginning of an inning, the expected number of runs scored is 0.47. If the first batter gets a double, there is now a runner on second base with no outs, and the expected number of runs scored is 1.11. One measures the runs value of this double by computing the difference in expected runs scored.

$$RUNS = 1.11 - 0.47 = 0.64$$

This double contributes, on average, about a half of run scoring for the team.

THE RUNS EXPECTANCY MATRIX

Above the expected runs was computed for two situations, no runners on with no outs, and a runner on second base with no outs. Suppose one repeats the calculation for each possible situation of runners on base and number of outs. Each of the three bases (1st, 2nd, and 3rd) can be occupied or not, so there are $2 \times 2 \times 2 = 8$ possible configurations of runners on base. Also there can be 0, 1, or 2 outs, so there are $8 \times 3 = 24$ possible combinations of runners of base and outs.

For each scenario, one computes the average or expected number of runs in the remainder of the inning. These numbers represent the potential to score runs in different runner and outs situations, so they are sometimes called *runs potentials*. When these average runs values are presented in matrix form where the rows correspond to the number of outs and the columns correspond to the runner configurations, this is called the runs expectancy matrix. This matrix for 2015 season data is displayed in Table 3.1.

Table 3.1 The runs expectancy matrix based on data from the 2015 season.

RUNNERS	000	100	020	003	120	103	023	123
OUTS = 0	0.47	0.86	1.11	1.40	1.47	1.71	2.04	2.29
OUTS = 1	0.25	0.50	0.66	0.96	0.89	1.12	1.37	1.59
OUTS = 2	0.10	0.23	0.30	0.36	0.43	0.45	0.56	0.79

Figure 3.3 gives a graphical representation of the values in the runs expectancy matrix. The horizontal scale is the bases occupied, the vertical scale is the expected number of runs scored, and the plotting symbol gives the number of outs.

Some general patterns in these runs expectancies can be seen from this graph.

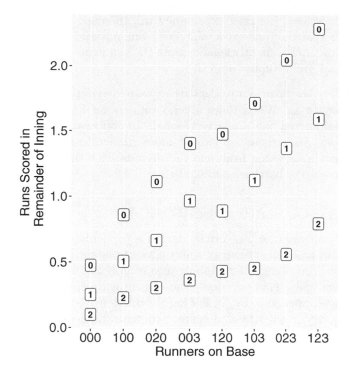

Figure 3.3 A graph of the values of the runs expectancy matrix for the 2015 season. The runners on base scale on the horizontal axis is ordered from fewer runners to more runners and the plotting symbol gives the number of outs.

- The situation with greatest run potential is the bases loaded with no outs. At the other extreme, very few runs will be scored when the bases are empty with two outs.

- For a given configuration of runners on base, say runners on first and second base ("120"), the runs expectancy is greatest with no outs, followed by one out, and then two outs. If an out occurs without changing the position of the runners, the runs expectancy will always go down. In other words, from the viewpoint of the batting team, an out costs the team a specific number of runs.

- The runner positions on base are arranged in the graph from left to right as "no runners" ("000"), "1 runner" ("100", "020", "003"), "2 runners" ("120", "103", "023"), and "3 runners" ("123"). Generally, as the number of runners on base increases, the runs expectancy will increase. One exception is that the runs expectancy with one out and runners on 1st and 2nd is slightly smaller than the run expectancy with one out and a runner on 3rd.

- The cost of an out to the batting team depends on the runners situation. When there are no runners on base, the cost of adding an out is relatively small. In contrast, the cost of an out when there are two or more runners can be high. For example, going from one to two outs with the bases loaded costs the batting team 0.80 runs.

VALUES OF PLATE APPEARANCES

A plate appearance is a turn at-bat for a particular player. A half-inning of baseball consists of a sequence of plate appearances, and the outcome of each plate appearance usually has an effect on the (runners, outs) configuration. One can measure the runs value of the plate appearance by (1) finding the difference between the runs expectancy in the new and current configurations and (2) adding any runs scored on the play. As a formula

$$Value = RUNS_{new} - RUNS_{current} + (RUNS\ SCORED)$$

Example 1: Result of a strikeout

As a simple example, suppose there is a runner on first base with no outs – by the graph (Figure 3.4), the runs expectancy is 0.86 runs.

The batter strikes out and the new state is runner on first base with one out. This transition is displayed by the blue arrow in Figure 3.4. The runs expectancy of this new state is 0.50 runs and no runs scored on this play. So the value of this strikeout is

$$Value = 0.50 - 0.86 + 0 = -0.36$$

As expected, this strikeout has a negative runs value from the batting team's perspective.

Example 2: A single scoring a runner, and a second runner thrown out at home

Next, suppose there are runners on second and third bases with one out. The batter gets a single, scoring the runner from third, but the runner on second base gets thrown out at home. This movement in

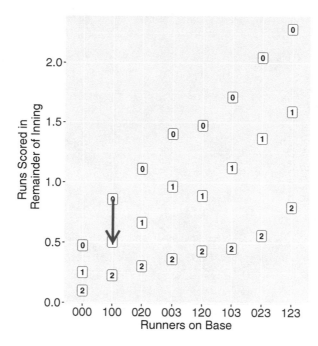

Figure 3.4 Change in runs expectancy when there is a strikeout when a runner is on 1st base with 0 outs.

states is shown in Figure 3.5 by a blue arrow in the run potential graph.

The runs expectancy of the new state (runner on first with two outs) is 0.23 runs and the runs expectancy of the current state is 1.37. One run was scored on the play. Here the value is

$$Value = 0.23 - 1.37 + 1 = -0.14$$

In this case, the value of the play to the batting team is close to zero. The run that is scored on the play is offset by the runner thrown out at home.

Example 3: A double with the bases loaded

As a final example, suppose the bases are loaded with no outs. The batter gets a double, clearing the bases so three runs are scored. This transition is shown graphically in Figure 3.6 by a blue arrow in this graph. What is the value of this double in this situation?

The runs potential of the bases loaded state with no outs is 2.29 runs. After the double, there is a runner on second base with

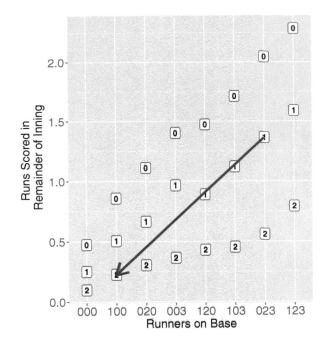

Figure 3.5 Change in runs expectancy with a single with runners on 2nd and 3rd bases with one out, and the runner on 2nd thrown out at home.

no outs which has a potential of 1.11 runs. Recall that three runs were scored on the play. So the value of this bases-loaded double is

$$Value = 1.11 - 2.29 + 3 = 1.82$$

As expected, this double had a relatively large value, but it is smaller than the actual number of runners who scored, since the new runners state (a runner on second base) is less valuable than the current runners state of bases loaded.

OTHER USES OF RUNS EXPECTANCY

There are many applications of runs expectancy in baseball and several of these applications are illustrated by use of graphs.

Values of Outcomes of a Plate Appearance

In a plate appearance, there are four types of hits: a single, double, triple, and home run. Although one typically attributes 1, 2, 3, and

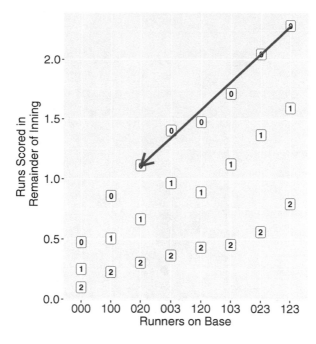

Figure 3.6 Change in runs expectancy with a double with the bases loaded where all runners on base score.

4 bases to these hits, it is unclear if these base numbers correspond to the value of these types of hits from a runs perspective. How does the value of a walk compare with the value of a single? Also, from a batter's perspective, there are negative outcomes such as a strikeout and an in-play out. Do these types of outs have the same value?

Figure 3.7 graphs the average runs value (over all possible runner and outs situations) for all of the different outcomes of a plate appearance. A vertical line at the value zero is placed on the plot so one can clearly distinguish positive and negative hitter results. The size of the plotting point corresponds to the percentage of plate appearances in the different outcomes.

There are several interesting observations from viewing this figure.

- The average runs values of a single, double, triple, and home run are respectively 0.44, 0.74, 1.03, and 1.37. In terms of bases, a double is twice as valuable as a single, but in terms of runs, a double is only $0.74/0.44 = 1.68$ times as valuable as a single. Using similar calculations, a triple is 2.34 times as

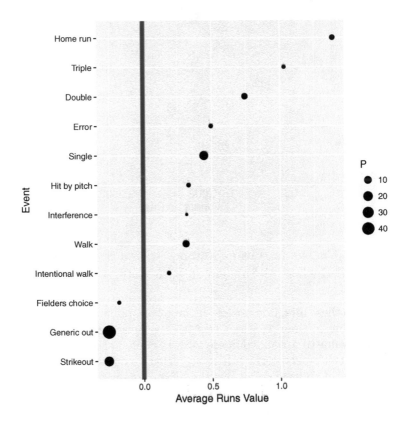

Figure 3.7 Average runs value for all possible outcomes of a plate appearance. The sizes of the plotting point corresponds to the percentages of plate appearances in the different outcomes.

valuable as a single, and home run is 3.16 times as valuable as a single.

- An in-play out and a strikeout are the same from a runs perspective – the value of each is −0.26 runs.

- Walks are not all equivalent. An intentional walk is worth 0.18 runs and an unintentional walk is worth 0.33 runs. This is reasonable, since an intentional walk is usually given when there is an open base and filling the open base will have a smaller runs value than a walk that advances the runners.

Values of a Home Run

Runs expectancies can also help measure the value of particular types of base hits in different situations. Figure 3.8 displays the value of a home run in each of the 24 bases/outs situations. The vertical line is drawn at the overall mean runs value of a home run. The size of the plotting point indicates the percentage of home runs in each situation.

Many of the Major League home runs are hit with no runners on base. In this case, the runners and outs situation will be the same before and after the home run and the runs value is exactly one. In contrast, the runs value of any home run with runners on base will exceed one. As one can see from Figure 3.8, the most valuable home run is the one hit when the bases are loaded with two outs – the runs value of this home run exceeds three.

WRAP-UP

This chapter provides a graphical view of runs expectancy which is one of the building blocks for understanding the values of baseball plays. A situation during an inning is defined by the runners and number of outs and a runs expectancy measures the potential to score runs in that situation. At the beginning of the inning approximately half a run, on average, is scored in the inning. Every batting play causes a possible change in the game situation and runs expectancy is used to measure the value of the play.

Runs values of plays have many applications in baseball analytics. They can be used to measure the contribution of specific plays such as a walk or a triple, they can be used to measure performance of players, and they can be used to examine baseball strategy as deciding whether to steal a base. This important baseball concept is introduced early in this book, since several applications of runs expectancy will be seen in future chapters. In particular, Chapter 4 on the count will use runs expectancy to measure the value of each additional pitch of a strike or a ball.

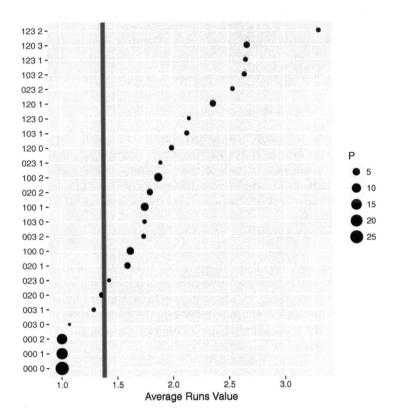

Figure 3.8 Value of a home run in all possible runner and outs situations. The sizes of the plotting points correspond to the percentages of home runs in the different situations. The location of the vertical line is at the mean runs value of a home run.

CHAPTER 4

The Count

INTRODUCTION

A plate appearance in a baseball game consists of a series of pitches thrown by the pitcher to the batter. The plate appearance ends when there is a strikeout, a walk, a hit-by-pitch, or the batter puts the pitch in-play. The count is the record of balls and strikes in the plate appearance. The dynamics between the pitcher and batter changes with every new pitch. The ball and strike count tells us, on face value, if the pitcher or batter has an advantage. For example, the pitcher has an advantage with a count of two strikes since three strikes are a strikeout, and a count with three balls favors the batter since four balls are a walk.

In Chapter 3, it was seen that expected runs is useful for determining the value of plays and this can be used to measure the value of hits in different situations. In this chapter, the use of expected runs is explored to measure the pitcher and batter advantages in the count. We look at the runs values that go through particular counts, and also look at the values of in-play events on different counts. This use of expected runs will help us measure what it means to be ahead or behind in the count.

DESCRIPTION OF THE COUNT

A Plate Appearance

A basic component of a game of baseball is the plate appearance (PA) – the confrontation between a pitcher and a batter. The pitcher's objective is to produce an out for his team by a strikeout or an out on a ball in play. In contrast, the batter's objective is to get on base by a hit or a walk, or to produce runs by a home run or a hit advancing runners to home.

A PA consists of a series of pitches. Each pitch is recorded as a ball or a strike. A pitch not swung at by the batter landing outside of the (strike) zone is called a "ball." A pitch not swung by the batter falling inside the zone, or a pitch missed by the batter, or a pitch hit by the batter in foul territory is called a "strike." The PA ends when there are four balls (the batter gets a walk), if there are three strikes (the batter strikes out), or the ball is hit by the batter in fair territory (the ball is "in-play"). One caveat is that a strikeout is recorded only when the final pitch is a called strike or a swinging strike.

The Count

The *count* is a record of the number of balls and number of strikes in the PA. One records the count as "b-s", where b is the number of balls and s is the number of strikes. The PA begins with a 0-0 count – if the first pitch is a strike, the count changes to 0-1 and if the first pitch is a ball, it changes to 1-0. The 12 possible counts in a PA are

0-0, 1-0, 0-1, 2-0, 1-1, 0-2, 3-0, 2-1, 1-2, 3-1, 2-2, 3-2

A PA basically is a movement through different counts until there is a walk (or hit-by-pitch), a strikeout, or a ball put in-play. If there is a foul on a two-strike count, say 1-2, the count will remain at 1-2. It is possible that the count will remain at two strikes for many pitches if the batter keeps hitting foul balls.

Some Sample Plate Appearances

To illustrate sequences of counts, Table 4.1 displays pitches thrown in the first three PAs in the 2016 World Series Game 7 between the Cleveland Indians and the Chicago Cubs. The table shows the pitcher (Corey Kluber of the Indians), the Cubs hitters (Dexter Fowler, Kyle Schwarber, and Kris Bryant), the current count, and the outcome of the pitch (ball or strike). Note how the count changes as a result of the outcome on the previous pitch.

Table 4.1 Pitches to the first three batters in Game 1 of the 2016 World Series.

Pitch #	Pitcher	Batter	Count	Outcome
1	Kluber	Fowler	0-0	Called Strike
2	Kluber	Fowler	0-1	Ball
3	Kluber	Fowler	1-1	Ball
4	Kluber	Fowler	2-1	In play, home run
5	Kluber	Schwarber	0-0	Ball
6	Kluber	Schwarber	1-0	Foul
7	Kluber	Schwarber	1-1	Called Strike
8	Kluber	Schwarber	1-2	In play, single
9	Kluber	Bryant	0-0	Swinging Strike
10	Kluber	Bryant	0-1	Swinging Strike
11	Kluber	Bryant	0-2	Ball
12	Kluber	Bryant	1-2	Ball In Dirt
13	Kluber	Bryant	2-2	Ball
14	Kluber	Bryant	3-2	In play, flyout

Pitcher and Hitter Counts

At the beginning of the PA, a 0-0 count, there is no special advantage to the pitcher or the batter, but that will change as pitches are thrown. The pitcher would like to throw strikes with the desire of striking out the batter with three strikes. In contrast, the batter likes to see balls since there is the possibility of walking when there are four balls.

Particular counts during a PA are called "pitchers' counts" since they give an advantage to the pitcher. A count with two strikes such as 1-2 is an example of a pitchers' count. In this situation, the pitcher does not have to throw a strike and he can purposely throw a pitch outside of the zone that is difficult to hit. From the batter's perspective, he is concerned about striking out on the next pitch and he may change his batting approach, such as shortening his swing, to make it more likely to get the bat on the ball.

A count with many balls such as a 3-1 count is called a "hitters' count." In this situation, the pitcher is reluctant to throw a ball and walk the batter, and will try hard to throw a strike on the next pitch. On the other side, the batter is aware that the pitcher is trying to throw a strike, and expects to see the next pitch in the zone. When the batter knows the location of the pitch, it is more likely that he will get a productive swing on the pitch.

Generally, two strike counts such as 0-2, 1-2, and 2-2 are believed to be pitchers' counts, and high-ball counts like 2-0, 3-0, and 3-1 are viewed as hitters' counts. Categorization of the remaining counts such as 1-0, 0-1, 1-1, 2-1, and 3-2 is less clear and may be viewed as "neutral" counts neither favoring the pitcher nor the batter. Based on the above comments, one can order the counts as follows where the left side favors the pitcher and the right side favors the hitter (parentheses are used to group counts where there is not a clear ordering):

0-2, 1-2, 2-2, (0-0, 1-0, 0-1, 1-1, 2-1, 3-2), (2-0, 3-1), 3-0
Favors Pitcher   Favors Hitter

This discussion on counts raises several questions:

- Do the counts 1-0, 0-1, 1-1, 2-1, and 3-2 favor the hitter or the pitcher?

- Adding a strike to the count will benefit the pitcher. But are all additions of one strike equally advantageous to the pitcher? That is, is the change from 0-0 to 0-1 the same as a change from 0-1 to 0-2?

- Likewise, one knows one additional ball benefits the hitter. But is a change from 1-0 to 2-0 the same as a change from 2-0 to 3-0, or a change from 1-1 to 2-1?

- Generally, can one confirm the ordering of counts above by use of some numerical measure?

RUNS EXPECTANCY

Introduction

The questions about the pitcher or hitter advantage in particular counts can be addressed using the notion of runs expectancy discussed in Chapter 3. A batter comes to bat during a particular inning situation defined by the number of outs and runners on base. Using the runs expectancy table, one finds the average or expected runs scored in the remainder of the inning. At the end of the PA, the inning situation will change (maybe there will be new runners or perhaps there is an additional out), and one finds the average runs scored in the remainder of the inning for the new situation. The runs value of that PA is defined by

$$RUNS_{after} - RUNS_{before} + (Runs\ Scored\ on\ Play),$$

where $RUNS_{before}$ and $RUNS_{after}$ denote the average runs scored in the remainder of the inning before and after the PA.

Every PA has an associated runs value. If a batter gets out by a strikeout or an in-play out, the runs value will be negative and positive events such as a hit or a walk will correspond to a positive runs value.

One can extend the notion of runs expectancy to counts during a PA. As an example, suppose one wants to find the runs expectancy for a 2-0 count. One examines all of the PAs that pass through a 2-0 count, and find the average of the runs scored in the remainder of the inning for those PAs. This average is called the runs expectancy of a 2-0 count. Likewise, one can find the runs expectancy of any possible count.

A Graph of Run Expectancies of Counts

This runs expectancy calculation was performed for each of the 12 possible counts and Figure 4.1 constructs a graph of the runs expectancies against the pitch number. Lines are drawn between the labeled points to show the possible transitions of the count during a PA. For example, a 0-0 count can move to a 0-1 count to a 1-0 count, a 1-1 count can move to a 2-1 count or a 1-2 count, and so on. A horizontal line is drawn at the value 0. Points above the line correspond to hitters' counts with positive runs expectancies and points below the line correspond to pitchers' counts with negative runs expectancies.

What do we learn from Figure 4.1?

- Runs expectancy divides the counts into 6 positive counts (1-0, 2-0, 2-1, 3-0, 3-1, 3-2) and 5 negative counts (0-1, 1-1, 0-2, 1-2, 2-2), and one neutral count (0-0).

- These calculations help to distinguish the "in-between" counts 1-0, 0-1, 1-1, 2-1, 3-2 where the advantage to the pitcher or the hitter is uncertain.

- The graph shows the most extreme hitters' count is 3-0 and the most extreme pitchers' count is 0-2. This makes sense since many of these PAs with these counts lead respectively to walks and strikeouts.

- Adding a single strike to the count seems to drop the runs value by about 0.005, but the cost of a strike seems slightly larger for more advanced counts like 2-1 and 3-1.

- In contrast, the cost of a single ball seems to depend more on the current count. For example, the cost of a ball at 0-0 (moving to 1-0) seems small, but the cost of a ball at 2-0 (moving to 3-0) seems much larger.

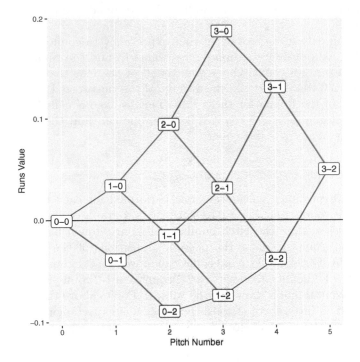

Figure 4.1 Graph of runs expectancies of PAs against the pitch number passing through each possible count.

- The graph is nice in that one can see easily how the pitcher/batter advantages can change within a single PA. For example, it is easy to imagine the sequence 0-0 to 1-0 (hitter advantage) to 1-1 (slight pitcher advantage) to 2-1 (hitter advantage) to 2-2 (pitcher advantage).

Run Expectancies of In-Play Events

Figure 4.1 shows how the pitcher or the hitter gets an advantage as one passes through different counts in a single PA. A different perspective is to look how the quality of a batted ball put in play depends on the current count. Based on observation of many baseball games, I believe that batters have a clear hitting advantage when the count is in their favor. For example, I believe many home runs are hit in hitters' counts where the batter is anticipating a pitch in the zone. In contrast, when there is a pitchers' count, the batter often takes a defensive or short swing that can result in a weak batted ball put into play.

One can measure the quality of a batted ball put into play by the use of runs expectancy. For example, for a 2-1 count, we look at all the PAs in the 2015 season where a ball was placed in play on a 2-1 count, and we compute the average runs scored in the remainder of the inning of these PAs. One repeats this calculation for each of the 12 possible counts.

Figure 4.2 displays these "in-play" runs values as a function of the pitch number. The points in this graph are not connected since the different plotting points are based on data from different PAs. The sizes of the plotting labels are proportional to the frequencies of batted balls in those cases. So, for example, we see that many balls are put in play on 0-0 and 0-1 counts, and it is uncommon to put a ball in play on 2-0, 3-1, and 3-0 counts.

What does one learn from this graph?

- All of the runs expectancies are positive indicating that it is advantageous to the batter to put the ball in play.

- The runs expectancies are smallest for the two-strike counts (0-2, 1-2, 2-2) and the 0-1 count.

- The runs expectancies for the 0-0, 1-0, 1-1, 2-1, and 3-2 counts are similar, although batters appear to perform better when the number of balls in the count exceeds the number of strikes.

- The in-play runs expectancies are highest for the hitter advantage counts 2-0, 3-0, and 3-1. But note that the size of the plotting symbol is small for these counts, which indicates that batters are less apt to put a ball in play when the count is in their advantage.

Count Effects for Individual Batters and Pitchers

In this chapter, we have focused on average count effects for all PAs in a particular season. But it is interesting to explore how these count effects vary between pitchers and between batters. In the following graphs we focus on runs expectancies for PAs that pass through all of the possible counts.

Figure 4.3 compares the count runs expectancies for two hitters, Albert Pujols and Mike Trout, teammates on the 2015 Los Angeles Angels.

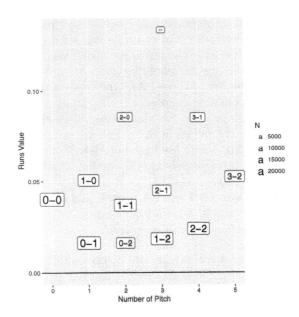

Figure 4.2 Graph of runs expectancies of in-play events for each possible count. The sizes of the plotting symbols correspond to the frequencies of these in-play counts for balls in play.

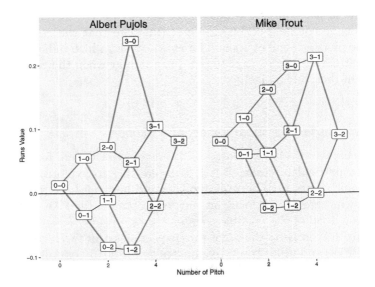

Figure 4.3 Runs expectancy graphs of counts for Albert Pujols and Mike Trout in the 2015 season.

In this particular season, Trout was one of the best hitters in the major leagues creating 8.9 offensive wins for his team. In contrast, Pujols had a relatively weak hitting season, creating only 2.6 offensive wins. As one might expect, the two hitters have very different count profiles. Pujols had a negative runs expectancy for all of the pitchers' counts, and had modest runs values for most of the other counts with the exception of the 3-0 count. In contrast, Trout had negative runs expectancy for only the 0-2 and 1-2 counts and really had large runs expectancy for the group of hitters' counts. Graphs like these are helpful in seeing how batters react to different counts during a PA.

These graphs can also be used to compare pitchers. Dallas Keuchel and Max Scherzer were two of the best pitchers in the 2015 season. The Baseball Reference site indicates that the WAR measures for these two pitchers were 7.2 and 7.1, respectively, indicating that they performed similarly that season. Figure 4.4 displays count runs expectancies for the two pitchers.

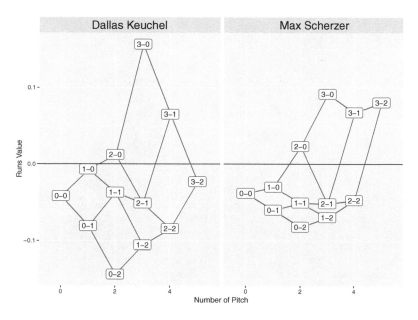

Figure 4.4 Runs expectancy graph of counts for Dallas Keuchel and Max Scherzer in the 2015 season.

If one looks at the runs value for count 0-0, the two pitchers are very similar, indicating that they performed the same across all PAs. But the two pitchers differ in how they perform on specific counts. For example, Keuchel's runs values for the two-strike counts (0-2, 1-2, 2-2) are smaller than Scherzer, indicating that Keuchel was better in taking advantage of these pitchers' counts. On the other hand, Scherzer had a smaller runs value than Keuchel for the 1-0 count, which says that Scherzer better handles PAs where the first pitch is a ball.

WRAP-UP

In this chapter, the concept of runs expectancy is extended to what happens in a single PA. Each pitch in a PA has an associated count and one can measure the value of the count by a runs expectancy. These runs expectancy calculations help to clarify the counts that favor the pitcher, the counts that favor the hitter, and order the value of the neutral counts. These type of calculations for specific batters and pitchers help in determining counts where these players tend to be successful. Graphs of these runs expectancy measures help us to better understand the dynamics in the pitcher-batter matchup that is an important component of a baseball game.

CHAPTER 5

PITCHf/x Data

INTRODUCTION

During the 2007 baseball season, Major League Baseball (MLB) began a systematic effort to record detailed information about the pitches that are thrown. All baseball stadiums were equipped with video cameras that would track each pitched ball and determine its precise trajectory. From the measurements made from the cameras, one is able to learn about the speed of each pitch at its release point and at the point where it reaches home plate. Also one can measure the amount and angle of the movement of the pitch in the path from the pitcher's release point to crossing the plate. This technology is known as the PITCHf/x system. A good introduction to the PITCHf/x system and the associated variables is given in Fast (2010).

By scraping data from the PITCHf/x system, one can learn quite a bit about the assortment of pitches thrown in Major League Baseball. One can learn about the pitches thrown by specific pitchers and find out why they are successful (or not successful) in getting batters out. In this chapter, we provide a gentle introduction to the rich PITCHf/x data, using graphs to explore tendencies of particular pitchers.

WHAT PITCHES DO THEY THROW?

There are many types of pitches thrown in modern professional baseball and these pitch types differ in terms of the speed that they are thrown and the movement or direction as they travel from the pitcher's hand toward the batter. In the PITCHf/x system, measurements are made on the velocity and movement in the horizontal and vertical directions, and these measurements are used to classify each pitch as a specific type. We learn about different pitch types by looking at four pitchers who had remarkable 2016 seasons.

Practically all pitchers in Major League Baseball throw more than one type of pitch. Figure 5.1 displays the percentage of pitches thrown of different types for the pitchers Chris Sale, Clayton Kershaw, Cole Hamels, and Jake Arrieta. Let's look at the pitch repertoire for each pitcher – in this exploration, we will describe the different pitch types.

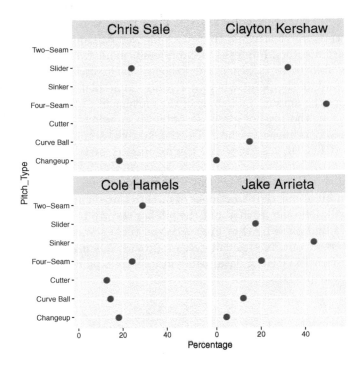

Figure 5.1 Percentages of different pitch types thrown by four starting pitchers in the 2016 season.

- Chris Sale throws three types of pitches – a two-seam fast-ball, a slider, and a changeup. A two-seam fastball is one type of fastball where two fingers are placed along a seam of the baseball. A slider is a slower type of pitch that moves laterally down through the batter's hitting zone. A changeup is a second type of slower pitch that is designed to look like a fastball but moves at a slower speed. One sees from the graph that about half of Sale's pitches are two-seam fastballs, and sliders and changeups are each thrown about a quarter of the time.

- Clayton Kershaw throws four types of pitches, but most of them are sliders, four-seam fastballs, and curve balls. A four-seam fastball, the most-common type of fastball, is thrown by gripping the ball across the wide part of the seam. A curve ball is one of the slower type of pitches that dives in a downward path as it approaches the batter. We see that Kershaw throws about 50% four-seamers, 30% sliders, and a smaller percentage of curve balls.

- Cole Hamels actually throws five distinct types of pitches, two-seam fastballs, four-seam fastballs, cutters, changeups, and curve balls. A cutter is a special type of fastball with movement and speed that is between a slider and a typical fastball. About half of Hamels' pitches are either two-seam or four-seam fastballs, followed by a changeup, a curve ball, and a cutter.

- Jake Arrieta also throws five pitches, a sinker, a slider, a four-seam fastball, a curve ball, and a changeup. A sinker is a special type of fastball with a significant downward move-ment. One sees from Figure 5.1 that almost half of Arrieta's pitches are sinkers, followed by sliders, two-seam fastballs, curve balls, and changeups.

Some general statements can be made about pitch selection by exploring the pitch types of these four pitchers. First pitchers appear to throw multiple types of pitches, and each pitcher throws some type of fastball (such as a two-seamer, a four-seamer, a cutter, or sinker) and some type of "off-speed" pitch such as a curve ball, changeup, or slider.

PITCH VARIABLES

We next discuss three key attributes of a pitch – its speed, its movement, and its location.

Pitch Speed

To help distinguish these pitch types, we focus on the pitches thrown by Clayton Kershaw. Figure 5.2 displays parallel boxplots of the pitch speeds of the four types of pitches. This graph tells us there are clear distinctions between the pitch speeds of Kershaw's sliders, four-seam fastballs, and curve balls. The four-seamers are the fastest – the graph tells us that his four-seamers are consistently thrown close to 93 miles per hour. The sliders, in contrast, average about 88 mph, and the curve balls only 73 mph. If a batter is expecting Kershaw to throw a four-seamer and actually a curve ball is thrown, one can see how the batter is fooled. The curve ball arrives much slower to the plate than the fastball and it would be easy for the batter to swing too soon.

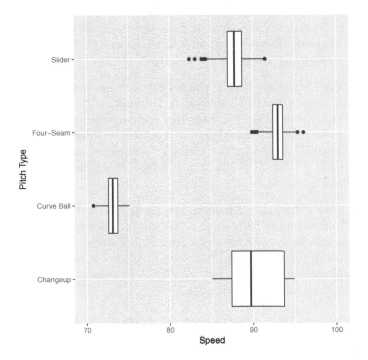

Figure 5.2 Boxplots of speeds of pitches of different types thrown by Clayton Kershaw in the 2016 season.

Pitch Movement

Besides speed, one can distinguish these pitch types by the movement of the ball as it travels from the pitcher to the batter. There are two variables in the PITCHf/x dataset that measure movement. The variable `pfx_x` is the horizontal movement (measured in inches) of the pitch from the pitcher's release point and home plate, as compared to a similar pitch thrown at the same speed without any movement. The variable `pfx_z` measures the vertical movement in inches from the release point to home point. The perspective of the movement measurements is from the catcher behind home plate. Figure 5.3 displays a scatterplot of the horizontal and vertical movements of Kershaw's pitches where the shape of the plotting point corresponds to the pitch type.

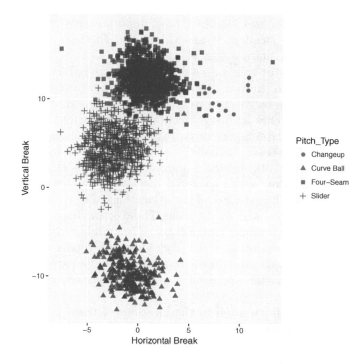

Figure 5.3 Horizontal and vertical movements (in inches) of pitches thrown by Clayton Kershaw in the 2016 season. Each different type of pitch is represented using a different plotting symbol.

The three clusters in the graph correspond to movements of the three pitch types. The four-seam fastballs generally have a larger vertical movement and little horizontal movement. The second cluster corresponds to sliders that have less vertical movement and a negative horizontal movement. A slider moves toward a right-handed batter and away from a left-handed hitter. The movement that identifies a curve ball is the strong negative vertical movement that drops the pitch below the lower edge of the strike zone.

Pitch Location

A baseball pitcher throws a pitch to the batter toward a "strike zone" or more commonly called "zone." This zone is a region (approximately a 17-inch square) that covers home plate and is between the midpoint of the batter's torso and the hollow beneath his knees. If the batter does not swing and the pitch lands within the zone, the pitch is a "called strike." If the pitch lands outside of the zone, it is called a "ball."

As in buying houses, the three most important aspects of effective pitching are location, location, location. Fastballs thrown in the middle of the zone are relatively easy to hit, and pitches thrown in the corners of the zone or outside the zone are harder to hit. A pitcher will try to throw a specific pitch, say a curve ball, toward a specific location inside or outside the zone and good pitchers are able to throw pitches with high accuracy to the desired location.

Figures 5.4 and 5.5 present two graphs of the locations of the curve balls, four-seam fastballs, and sliders thrown by Clayton Kershaw in the 2016 season. Figure 5.4 displays scatterplots of the pitch locations, and Figure 5.5 displays an alternative two-dimensional density estimate of these locations. The rectangle corresponds to the location of the zone. These estimates consist of a group of concentric circles where the smallest circle corresponds to the greatest concentration of the pitch locations. Again, remember that these locations are viewed from the catcher's perspective so a right-handed batter would be standing on the left side of this diagram.

There is much variation in the locations of these pitches, but some general patterns can be seen. Four-seam fastballs (upper right panel of the figures) tend to be thrown in the middle of the strike zone. Sliders (lower left panel) tend to fall lower in the strike zone toward a right-handed batter (and away from a left-handed hitter). This behavior is consistent with the direction of the horizontal movement of the slider. Last, curve balls (upper left panel) tend to fall low, actually outside of the zone. This is consistent with the sharp downward movement of the curve ball.

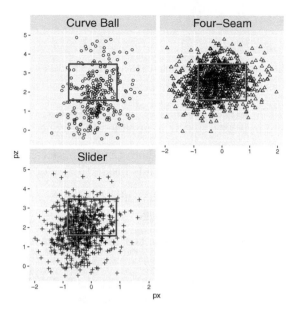

Figure 5.4 Scatterplot of locations of pitches thrown by Clayton Kershaw in the 2016 season for three pitch types.

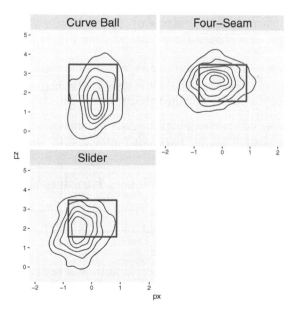

Figure 5.5 Density estimates of locations of pitches thrown by Clayton Kershaw in the 2016 season for three pitch types.

THE PITCH OUTCOME

Outcome of the Pitch

What is the outcome of the pitch thrown to the batter? We distinguish six basic outcomes:

- A *ball* is a pitch where the batter does not swing and the umpire indicates that it is outside of the zone.

- A *called strike* is a pitch where the batter doesn't swing and umpire indicates that it lands inside the zone.

- A *hit-by-pitch* is a pitch that hits the batter and the batter gets to advance to first base.

- A *swing-and-miss* is where the batter swings and makes no contact with the ball.

- A *foul* is where the batter swings at the ball and it is hit into foul territory (not inside the playing field).

- An *in-play* happens when the batter swings and places the ball inside the playing field.

Figure 5.6 shows the percentage of each of the six outcomes for each of Kershaw's three pitch types. This graph is informative about the reaction of the batters to each of Kershaw's pitches. Here are some observations from viewing Figure 5.6.

- A high percentage of Kershaw's curve balls are called "ball." This indicates that Kershaw is usually trying to get the batter to chase a curve ball thrown outside of the strike zone.

- Comparing the "swing-and-miss" for the three pitches, it is much more likely for batters to miss a slider than a fast ball or a curve ball.

- A called strike is more likely for a fastball than a curve ball or slider.

- The percentage of balls put in play is similar for all three pitches.

- It is more likely for a batter to foul off a fastball or a slider than a curve ball.

- It is rare for a batter to be hit by a pitch.

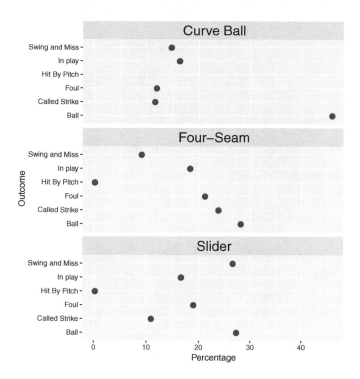

Figure 5.6 Percentages of pitches of different outcomes for each of the different pitch types of Clayton Kershaw in the 2016 season.

Outcome of the Swing

Next we focus on the pitches where the batter swings. In this case, there are three basic outcomes, either the batter misses the pitch, he fouls the pitch, or puts the ball in play. Figure 5.7 shows the percentages of these three outcomes of "swing" for each of the three pitch types thrown by Clayton Kershaw.

The message from Figure 5.7 is that a batter is more likely to miss a slider, and he rarely misses a four-seamer. The batter is more likely to put a curve ball or a fastball in play, and he is most likely to hit a foul on a fastball.

To gain a better understanding why batters swing-and-miss on particular pitch types, Figure 5.8 displays the locations of all swung pitches from Kershaw for the three pitch types. The points are colored by the outcome – a darker point corresponds to a swing that is missed.

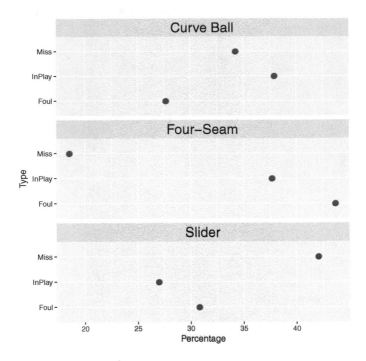

Figure 5.7 Percentages of swung pitches that are either missed, put in play, or fouled for each of the different pitch types of Clayton Kershaw in the 2016 season.

It can be seen from this graph that batters tend to miss fastballs outside of the zone, both high and left of the zone. Batters miss curve balls that are low, in fact a batter can be fooled by the pitch type and miss a curve ball that is far below the zone. It was earlier seen that it was most common for batters to miss sliders. Figure 5.8 tells us that these pitches tend to be missed in the lower-left region outside of the zone.

Quality of the Ball Put In Play

The batter's objective is to put the ball in play with a positive outcome such as a base hit. Particular base hits such as a double or a home run are especially valuable since these hits can advance runners on base to home plate to score runs. From a pitcher's perspective, he would like to throw a pitch that leads to a weak ball put in play likely to be an out. Here the outcomes are explored of pitches from Clayton Kershaw that are put into play. Figure 5.9

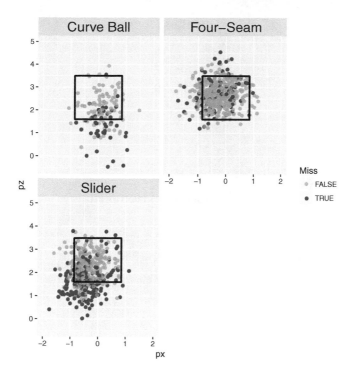

Figure 5.8 Locations of "swung" pitches thrown by Clayton Kershaw in the 2016 season for three pitch types. The result of the swing is indicated by the color of the plotting point.

displays the batting average and the fraction of home runs hit on balls put in play against Kershaw's pitches.

A typical batting average for a Major League player is about 0.270. Looking at the top panel of Figure 5.9, we see that a batter likes hitting a four-seamer   the batting average for four-seamers put into play exceeds 0.300. In contrast, batting average on slider or curve balls is in the smaller 0.210 to 0.240 range. From Kershaw's perspective, sliders and curve balls are preferable to fastballs since they lead to weaker balls put in play.

A similar pattern is true for home runs. Home runs off of Kershaw are pretty uncommon, the proportions displayed in the bottom panel range from 0.0175 to 0.0275. But it is much more likely for a batter to hit a home run on a fastball.

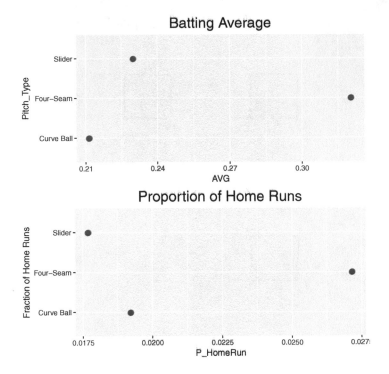

Figure 5.9 Batting average and fraction of home runs hit in balls in-play for each of Clayton Kershaw's pitch types for the 2016 season.

WRAP-UP

A major component of the game of baseball is the confrontation between the pitcher and the batter. The PITCHf/x data allows us to learn about the art of pitching. Each pitcher throws a variety of pitch types, and the effectiveness of a given pitch depends on its speed, movement, and location of the pitch about the zone. Great pitchers such as Clayton Kershaw throw several types of pitches that are noteworthy in terms of speed or movement, but more importantly he is able to precisely throw the pitches to different locations in the corners or just outside of the zone. Also for the pitchers who have unsuccessful appearances, a careful examination of the PITCHf/x data helps the manager and coaches understand what went wrong. As this type of data has been collected over 10 years, it will allow one to understand the changes in pitching over seasons.

CHAPTER 6

Batted Balls

INTRODUCTION

In many sport events, the size of the playing field is uniform. For example, American football is played in a rectangular field of dimensions 120 yards by 53.3 yards and the size of a professional basketball (NBA) game is 94 by 50 feet. Although the layout of the infield is the same for all Major League Baseball parks, there are notable differences in the physical dimensions of the outfield. In addition, there are many variables that influence the path of the home run such as the distance to the fence, the wind and temperature conditions, and the altitude of the park. We begin this chapter on batted balls by exploring the variability in the number of home runs hit in the 30 MLB ballparks.

Currently there are multiple variables collected on home runs, through the StatCast system (MLB Advanced Media, 2015), including the horizontal angle, the exit velocity, the elevation angle, and their distance. By exploring this data, we will look for interesting trends in home run direction, and how the direction can vary across ballparks. The relationship between exit velocity and distance for all home runs in the 2016 season will be explored, and this exploration will help determine favorable launch angles for hitting home runs.

More generally, the relationships between exit velocity, launch angle, and batted ball outcome are explored. What combinations of exit velocity and launch angle lead to batted balls that are likely to fall as base hits? This chapter is concluded by applying a statistical model that can predict the probability of a base hit based on the exit velocity and the launch angle.

HOME RUNS

Ballparks Effects

In Major League Baseball there are 30 teams and 30 ballparks, and each ballpark is known as one that is either home run friendly or one where home runs are harder to hit. To begin exploring the ballpark effects, Figure 6.1 displays the number of home runs hit in each of the 30 stadiums in the 2016 season. This graph tells us that AT & T Park (San Francisco Giants), Marlins Park (Florida Marlins), and Turner Field (Atlanta Braves) only yielded 120 to 130 home runs in 2016; in contrast, there were three parks, Safeco Field (Seattle Mariners), Yankee Stadium (New York Yankees), and Great American Ballpark (Cincinnati Reds) which surrendered over 220 home runs.

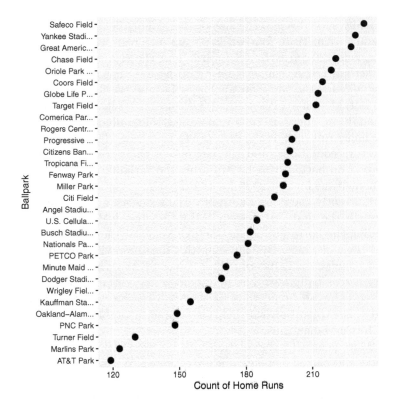

Figure 6.1 Number of home runs hit in each ballpark in the 2016 season.

Why do we see great differences in the numbers of home runs hit across ballparks? One possible explanation is the differences in the exact dimensions of the park and climate and weather factors such as wind, temperature, and altitude that can contribute to these home-run count differences. But perhaps these ballpark differences can be explained by the variation in the slugging ability of the home teams. Yankee Stadium might be home-run friendly since their lineup is stocked with home-run hitters.

One simple way of adjusting for the different team slugging abilities is to look at the difference in the total number of home runs hit by a team and their opponent for home and away games:

$$Home\ Runs\ at\ Home\ Games - Home\ Runs\ at\ Away\ Games$$

So in particular, for the Phillies, one computes the total number of home runs hit in Citizens Bank Park and subtracts the total number of home runs hit in the games where the Phillies were the visiting team. If these differences are computed for all teams, one gets the so-called "Park Factors". Figure 6.2 displays a scatterplot of the number of home runs against the park factors for all ballparks.

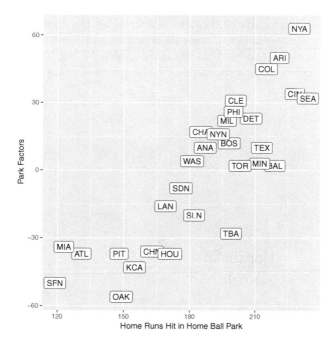

Figure 6.2 Scatterplot of ballpark home run counts against the park factors for all teams in the 2016 season.

As one would expect, there is a positive association in this graph. Ballparks allowing many home runs generally correspond to parks with positive park factors, and likewise home-run challenged ballparks correspond to negative park factors. But there are differences in the ordering of the ballparks using the park factor measure. For example, the easiest ballpark in which to hit home runs is Yankee Stadium according to park factors, and the hardest ballpark in which to hit home runs is Oakland Alameda Coliseum. (Recall that Safeco Field in Seattle and AT&T Park in San Francisco were the top and bottom ballparks with respect to total home runs hit.) Toronto actually has a park effect close to zero despite the fact that its ballpark Rogers Centre allowed a large number of home runs. Since one makes adjustments for team differences in home run hitting, the park factors are a better measure of ballparks' tendencies to influence the home runs.

What Directions are Home Runs Hit?

The ESPN Home Run Tracker provides the horizontal angle which is the initial angle of the ball as it leaves the bat. As shown in Figure 6.3, the horizontal angle is defined as the clockwise angle from the horizontal bat. Using this definition, an angle of 45 degrees is straight down the left field line, an angle of 90 degrees is straight over second base, and an angle of 135 degrees is straight down the right field line.

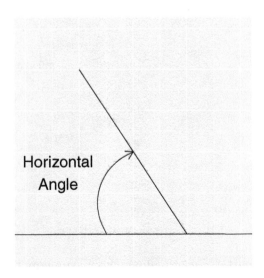

Figure 6.3 Definition of the horizontal angle of a home run.

Figure 6.4 displays a density curve of the horizontal angles using five years of home run tracker data for the 2012 through 2016 seasons. Note that all possible horizontal angles are not equally likely for home runs. Indeed, the most common angles are 70 and 110 degrees, corresponding to left and right field fences. It is less common for home runs to land in the middle of center field.

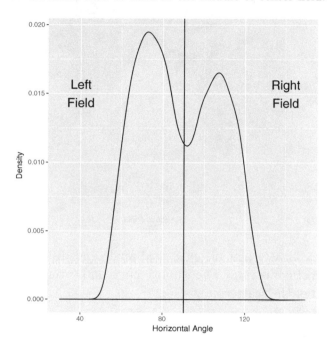

Figure 6.4 Density graph of the horizontal angles of all home runs from five seasons of MLB baseball.

Figure 6.5 explains why all horizontal angles for home runs are not equally likely by plotting the angle against the distance traveled for all home runs in our five-season dataset. The solid curve shows the general pattern in the scatterplot. Home runs hit to left and right fields have shorter distances than the home runs hit to center field. For example, a home run hit to mid left field (70 degrees) averages about 380 feet and a home run hit down the left field line (45 degrees) only averages about 360 feet. In contrast, home runs hit to center field (90 degrees) average about 420 feet. Many hard balls hit in the deepest part of the ballpark in center field are outs rather than home runs, and so it is relatively difficult to hit a home run with an angle close to 90 degrees.

The direction of home runs depends on the batting side of the hitter. Figure 6.6 displays density curves of the horizontal angles

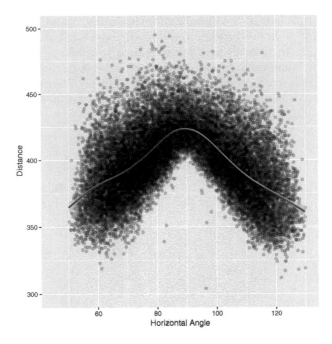

Figure 6.5 Scatterplot of horizontal angle and distance of five seasons of home runs. The smooth curve indicates the general pattern of the scatterplot.

of the home runs hit by left-handed and right-handed hitters from our five-season data. Generally batters pull their home runs which means that left-handed batters tend to hit home runs to right field and right-handed batters hit to left field. This pattern is confirmed in Figure 6.6 – the top panel tells us that left-handed hitters tend to hit home runs with a horizontal angle of 100–110 degrees (middle of right field), and right-handers have a mirror image pattern, hitting home runs with an angle of 70–80 degrees (middle of left field).

After seeing Figure 6.6, one can make more sense of Figure 6.4 that displayed a graph of the horizontal angles for all home runs. A majority of batters in current MLB are right-handed and these tend to hit to left field. For these reasons the direction of home runs will tend to favor left field instead of right, explaining the higher peak at 70 degrees in Figure 6.4.

Home Run Direction by Ballpark

It has already been established that there are ballpark effects with respect to home run hitting – it is easier to hit a home run in some

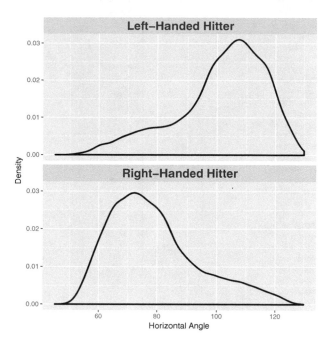

Figure 6.6 Density graphs of horizontal angles of home runs hit by left-handed (top) and right-handed (bottom) hitters.

ballparks than other parks. There are also ballpark effects with respect to the direction of the home runs that are hit.

In our five-year dataset, all home runs were divided into two groups – the ones hit toward left field (a horizontal angle less than 90 degrees) and the ones hit toward right field (an angle larger than 90 degrees). Figure 6.7 displays the fraction of home runs hit toward left field for all MLB ballparks. Generally, for most ballparks, the proportion of home runs hit toward left ranges between 0.5 and 0.6. But there are interesting outliers. Some ballparks appear to favor left-handed hitters and have a small proportion of home runs hit to left field, and other ballparks favor right-handers and have a high proportion of home runs hit to left.

To look more carefully at these ballpark direction effects, Figure 6.8 displays density curves of the horizontal angles for four "lefty favorite" ballparks and eight "righty favorite" ballparks. From viewing this graph, one sees that Progressive Field (Cleveland Indians), Yankees Stadium (New York Yankees), PNC Park (Pittsburgh Pirates), and Safeco Field (Seattle Mariners) all tend to favor left-handed batters for hitting home runs. Yankee Stadium has a famous reputation for its short 314 feet dimension to right field

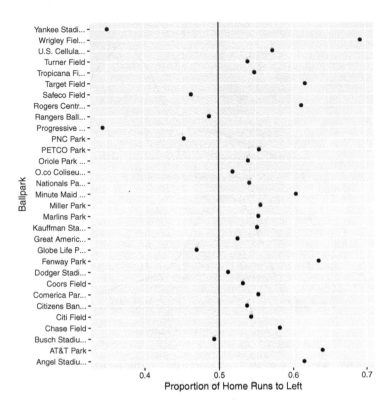

Figure 6.7 Proportion of home runs hit to left side of park for all MLB ballparks.

and many famous Yankee left-handed or switch-hitting sluggers such as Babe Ruth, Lou Gehrig, Roger Maris, and Mickey Mantle have benefited with this advantage. On the other side, one also sees from this graph that Wrigley Field (Chicago Cubs), AT & T Park (San Francisco Giants), Fenway Park (Boston Red Sox), Target Field (Minnesota Twins), Angel Stadium (Anaheim Angels), Rogers Centre (Toronto Blue Jays), Minute Maid Park (Houston Astros), and Chase Field (Arizona Diamondbacks) all tend to favor right-handed hitters as a greater proportion of home runs go to left field. Fenway Park also has a famous reputation for its "Green Monster" left field wall that lies only 310 feet from home plate.

Elevation Angle, Bat Speed, and Distance

Besides the distance traveled and the horizontal angle, the StatCast system collects the exit velocity, the speed of the ball off the bat

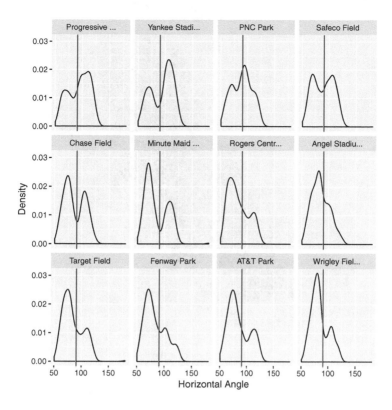

Figure 6.8 Density graphs of horizontal angles of home runs for 12 MLB ballparks. The four ballparks in the top row favor left-handed hitters and the remaining eight ballparks favor right-handed hitters.

(in mpg), and the launch angle (in degrees) for each ball placed in play. Figure 6.9 displays a scatterplot of the exit velocity and the distance traveled for all home runs hit in the 2016 season. Generally there is a positive association – harder hit home run balls tend to travel farther. But the strength of the association is not that strong. Suppose you look at all home runs hit at a particular exit velocity, say 100 mph. In the scatterplot this is represented by a vertical line at the value speed = 100. For these home runs with an exit velocity of 100 mpg, there is much variability in the distance traveled. So there are likely other variables besides bat speed that influence the distance traveled.

Another important variable in successful hitting is the launch angle, the vertical angle at which a ball leaves the bat. A launch angle of zero degrees corresponds to a grounder and certainly a

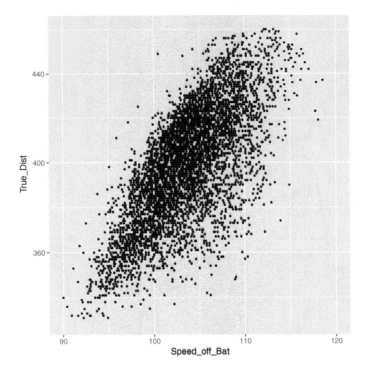

Figure 6.9 Scatterplot of exit velocity and distance traveled for all home runs hit in the 2016 season.

hitter wants a positive launch angle to have any chance of hitting a home run. But what are good values of launch angles toward the goal of getting a longer distance in a home run?

Figure 6.10 overlays two lines on the scatterplot of speed off the bat and distance. The solid line shows the relationship between bat speed and distance for balls hit with a launch angle between 20 and 30 degrees, and the dashed line shows the relationship for balls hit with an angle between 30 and 35 degrees. Since the dashed line falls over the solid line, this shows the superiority of a larger launch angle. For example, with a bat speed of 100 mpg, the graph tells us that a batted ball will travel 10 additional feet if the launch angle is 30–35 instead of 20–30 degrees.

BALLS PUT IN PLAY

StatCast: Exit Velocity and Launch Angle

In the exploration of home run data, we saw the relevance of the StatCast variables exit velocity and launch angle. Harder hit home

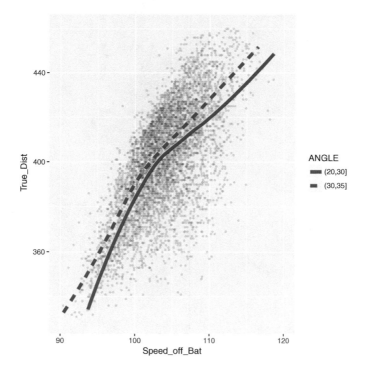

Figure 6.10 Scatterplot of exit velocity and distance traveled for all 2016 home runs. The smoothed curves show the relationships when the launch angle is between 20 and 30 degrees and when the launch angle is between 30 and 35 degrees.

runs tend to travel longer distances, and there are better choices for launch angle that will lead to home runs of longer distances. Here a large collection of balls put into play is explored. Knowing the exit velocity and launch angle, is it possible to make reasonable predictions of the probability the batted ball is a base hit?

We begin by looking at the data. Figure 6.11 shows a scatterplot of exit velocity and launch angle for 7296 batted balls for 21 batters in the 2015 season. A point is filled in if the particular batted ball is a base hit, and an open circle if the batted ball was an out. The vertical line corresponds to a zero value of a launch angle, so the points to the left of the line represent ground balls and points to the right represent balls hit in the air such as flyouts and popups.

Many of the points in the lower section of the scatterplot in Figure 6.11 correspond to softly hit balls that lead to outs. Figure 6.12 is a zoomed view of the scatterplot in Figure 6.11 where we focus on batted balls with an exit velocity of at least 50 mph and a

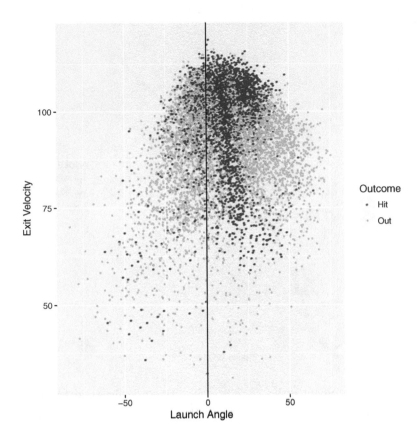

Figure 6.11 Scatterplot of exit velocity and launch angle for a large sample of batted balls where the outcome (hit or out) is indicated by the color of the point. The vertical line represents balls where the launch angle is zero.

launch angle exceeding −25 degrees. Generally it appears that hits are more likely for harder hit balls with a launch angle between 0 and 25 degrees. But the exact relationship between bat speed, launch angle, and hit/out is difficult to determine solely by a visual look at the data.

Modeling Hit Probabilities

A statistical model is helpful for understanding the relationship between exit velocity, launch angle, and the hit outcome. Let p denote the probability that a batter with a particular bat speed and launch angle obtains a base hit. A special type of model called

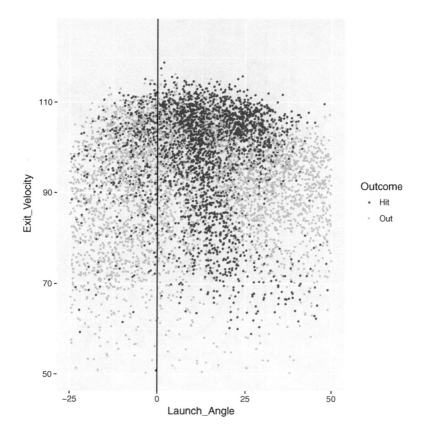

Figure 6.12 Zoomed view of a scatterplot of exit velocity and launch angle for a large sample of batted balls where the outcome (hit or out) is indicated by the color of the point.

a generalized additive model is applied where this hit probability is represented by a flexible function of the inputs bat speed and launch angle.

Figure 6.13 illustrates the application of the fitting of this generalized additive model to this batted ball data. The scatterplot of exit velocity and launch angle of the batted balls is presented from Figure 6.12. The lines overlaying the scatterplot represent the locations where the fitted probability of a hit is equal to the values 0.25, 0.50, and 0.75.

Figure 6.14 redraws this graph focusing on the predictions from the model. The darkest region represents the values of exit velocity and launch angle where the fitted probability of a hit exceeds 0.75, the next darkest region shows the values where the fitted

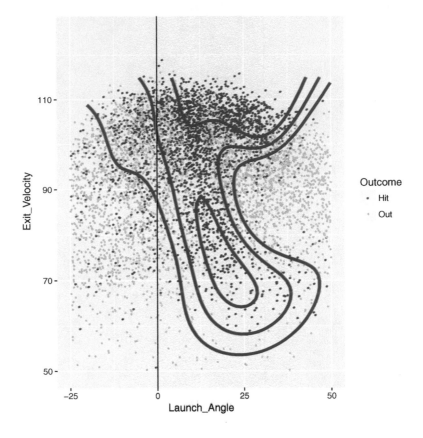

Figure 6.13 Scatterplot of exit velocity and launch angle for a sample of batted balls where the lines represent the fitted probabilities of a hit.

probability of a hit falls between 0.50 and 0.75, the lighter region corresponds to the area where the fitted probability falls between 0.25 and 0.50, and the white region is the area where the probability of a hit is smaller than 0.25.

There appear to be two distinct regions where the probability of a hit is the largest – the top region where the exit velocity exceeds 100 mpg and the launch angle is between 10 and 40 degrees, and a middle region where the bat speed is between 65 and 85 mpg and the launch angle is between 15 to 25 degrees. This tells us that base hits occur when the ball is struck very hard (think extra base hit such as a double or home run) or when the ball is hit at a moderate speed at an optimal launch angle. Outside of these two regions, there is a neighboring region where the probability of a hit is in the interval (0.50, 0.75). Generally, there are two types of outs (white region) – the groundballs that are not hit very hard and the flyball/popouts that are hit at moderate speed with a relatively high launch angle.

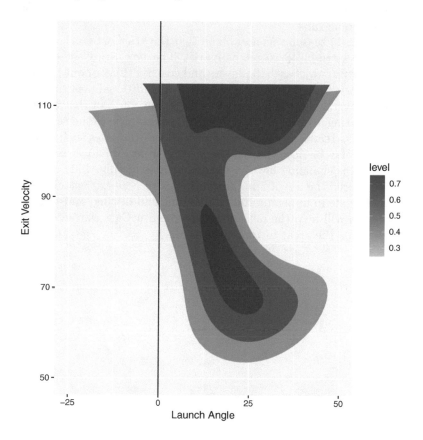

Figure 6.14 Graph of the fitted probabilities of a hit as a function of exit velocity and launch angle for a sample of batted balls. The three shaded regions represent the areas where the fitted probabilities exceed 0.75, fall between 0.5 and 0.75, fall between 0.25 and 0.50, and are lower than 0.25.

WRAP-UP

Ballparks play an important role in offensive performance in baseball. When one evaluates the hitting performance of a player, or compares the hitting performance of several players, one must take into account the ballpark where the players played. Todd Helton is currently considered for membership in the MLB Hall of Fame. Helton had remarkable hitting statistics in his career, but he played for many years in Coors Field, arguably the most hitter-friendly ballpark in the Major Leagues. One goal of this chapter is to explore ballpark effects with focus on the impact of the ballpark on hitting home runs.

The 2015 season was notable in that the StatCast tracking system was introduced to all ballparks. The new StatCast data is changing the way baseball players and coaches think about hitting. In the past, hitters were judged by use of traditional measures like batting average and slugging percentage, but much of the variability in batting average on balls put in play is not due to the variation in batters. Teams are learning that StatCast variables such as exit velocity may be better measures of batting performance. In short, StatCast is changing how baseball evaluates hitting. This chapter has explored the use of StatCast variables such as exit velocity and launch angle in understanding home runs and hitting, and this exploration will help the fan appreciate the StatCast characteristics that define the great hitters in baseball.

CHAPTER 7

Plate Discipline

INTRODUCTION

The FanGraphs website displays a large number of measures for baseball hitters and pitchers. If one looks at the Batting Leaderboard for a specific season, one sees a collection of tables titled "Standard", "Advanced", "Batted Ball", "Win Probability", "Pitch Type", "Pitch Value", "Plate Discipline", "Value", and "PITCHf/x". This chapter explores insights from the Plate Discipline table.

From a hitter's perspective, what is the meaning of plate discipline? First, the hitter needs to make an assessment of the location of the thrown pitch and decide whether to swing. Generally, a batter wants to swing at pitches located in the zone, and not swing at pitches thrown outside of the zone. As we will see in this chapter, the ability to not swing at pitches outside of the zone appears to vary greatly between hitters. Second, if the batter decides on swinging at the pitch, he would like to make contact. It will be seen that swinging rates and contact rates have much to do with strikeout rates and walk rates of batters.

The notion of plate discipline also has a meaning for pitchers. As we saw in Chapter 4, the pitcher wishes to gain an advantage in the count. To accomplish this, he wants to throw strikes and it is believed that it is especially important to throw a strike on the first pitch. So FanGraphs displays the percentage of first pitch strikes for all pitchers. We mentioned that batters wish to make contact with the pitch; conversely pitchers want the batter to swing and miss the pitch. So a second interesting pitching statistic is the percentage of swing and misses among all pitches thrown. As will be seen in this chapter, the first pitch strike rate and the swing and miss rate provide a reasonable prediction at a pitcher's wins above replacement (WAR) measure for a baseball season.

PLATE DISCIPLINE STATISTICS FOR BATTERS

Learning About Plate Discipline Statistics

The top right panel of Figure 7.1 shows the locations of 100 pitches thrown to Mike Trout during the 2016 season – 47 of these points were thrown in the zone and 53 were outside of the zone. The top right panel shows with a dark point the balls that were swung at – Trout swung at 25 balls in the zone, and 14 balls outside of the zone. The bottom left panel shows the balls where Trout made contact – here he made contact with 31 out of the 39 swings.

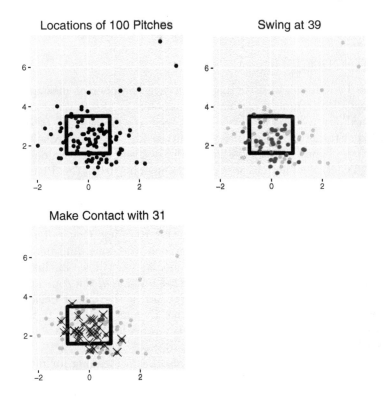

Figure 7.1 The top left figure shows the locations of 100 pitches thrown to Mike Trout during the 2016 season. The top right graph shows with a dark plotting point the balls that were swung at, and the bottom left graph indicates with an "X" the balls where Trout made contact.

Based on this data, FanGraphs defines seven "plate discipline" statistics:

- Zone% – the percentage of pitches thrown in the zone; here Zone% = 47/100 = 47%.

- Swing% – the percentage of pitches that were swung at; here Swing% = 39/100 = 39%.

- Z-Swing% – the percentage of pitches in the zone that were swung at; here Z-Swing% = 25/47 = 53%.

- O-Swing% – the percentage of pitches outside of the zone that were swung at; here O-Swing% = 14/53 = 26%.

- Contact% – the percentage of swings where there was contact; here Contact% = 31/39 = 79%.

- Z-Contact% – the percentage of swings on pitches in the zone where there was contact; here Z-Contact% = 22/25 = 88%.

- O-Contact% – the percentage of swings outside the zone where there was contact; here O-Contact% = 9/14 = 64%.

Exploring Swing and Contact Rates

Now that we have a basic understanding of some plate discipline measures, let's explore the swing and contact rates for all "qualifying" players in the 2016 season. Figure 7.2 displays a scatterplot of the swing and contact rates together with a smoothing curve to display the general pattern. One sees remarkable variation in the swing rates, ranging from 0.35 to 0.60. This means that some batters are reluctant to swing, swinging at only a third of the pitches, and other batters are "free swingers", swinging at over half of the pitches. Also, one sees much variation in the contact rates – some hitters successfully make contact with the bat 90% of the time, and other hitters struggle with a contact rate under 70%. Also, one sees a negative trend in the scatterplot, hitters with high swing rates tend to have small contact rates, and hitters who are more reluctant to swing tend to have higher contact rates.

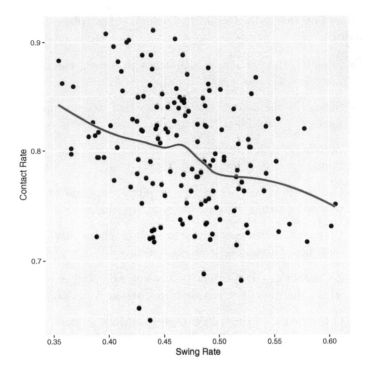

Figure 7.2 Scatterplot of swing and contact rates for regular players in the 2016 season. The smoothing curve indicates there is a negative association between these two rates.

In Chapter 1 it was seen that the current strikeout rate of baseball players is currently very high. How is a strikeout rate related to the swing and contact rates? This question is addressed by dividing the players into two groups – those with high strikeout rates (exceeding the median) and the remaining with low strikeout rates. Figure 7.3 redraws the scatterplot of swing and contact rates where the plotting point is colored by the strikeout group. Also, a line is drawn on the plot – this line represents the best discriminator between the high and low strikeout groups. The message from this figure is that the two groups are primarily distinguished by the contact rates – players with low contact rates tend to be more likely to strike out. Also, this figure indicates that there is a small "swing" effect. For a given contact rate, hitters with smaller swing rates are more likely to strike out.

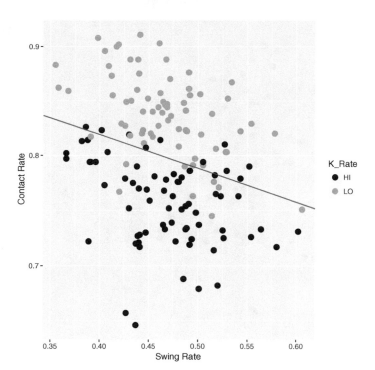

Figure 7.3 Scatterplot of swing and contact rates with a line representing the dividing line between players with high and low strikeout rates.

It is generally believed that batters with good plate discipline tend to draw more walks. Can we relate a batter's walk rate with his tendency to swing or make contact? Figure 7.4 again presents a scatterplot of swing and contact rates where the color of the plotting point represents batters in low and high walk rates. The line on the plot represents the best line for discriminating between the batters in the low and high walk groups. Note that the two groups tend to be distinguished by the swing rates – hitters with high swing rates tend to have small walk rates and hitters with low swing rates tend to have high walk rates. Interestingly, we also see a contact rate effect – for a given swing rate, hitters with low contact rates tend to walk more than hitters with high contact rates.

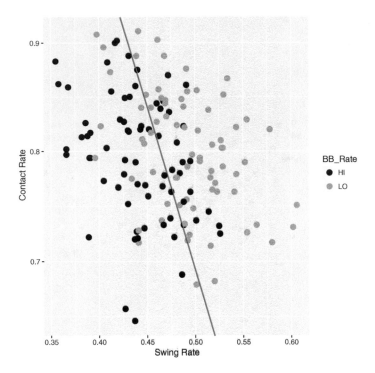

Figure 7.4 Scatterplot of swing and contact rates with a line representing the dividing line between players with high and low walk rates.

Rates in the Zone and Outside

We have focused on hitters' swing and contact rates and how they relate to strikeouts and walks. But what can one learn from the swing and contact rates for pitches in the zone and for pitches outside of the zone?

In our group of qualifying players for the 2016 season, approximately 15 (the top group) were identified who had the smallest strikeout rates, and 15 (the bottom group) who had the largest strikeout rates. Remember that strikeout rates have a strong association with contact rates. Figure 7.5 displays a scatterplot of the zone and outside contact rates for these top and bottom players. What one learns from this graph is that the top players (the ones who don't strikeout much) have better contact rates than the bottom players both for pitches in the zone and for pitches outside of the zone.

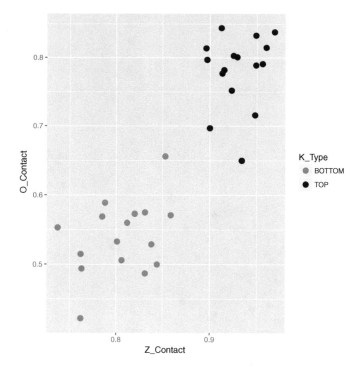

Figure 7.5 Zone and outside contact rates for players with low and high strikeout rates.

Figure 7.6 constructs a similar graph for walk rates. It was seen earlier that swing rates had a strong relationship with walk rates. A small group of players (the top group) is found with the highest walk rates and a second group (the bottom group) is found with the smallest walk rates, and Figure 7.6 displays a scatterplot of the zone and outside swing rates for the two groups. One sees from this graph that the two groups are clearly distinguished by their swing rates on pitches outside of the zone. The top group has outside swing rates under 30% and the bottom group has outside swing rates over 34%. The two groups are less distinguished by their swing rates on pitches in the zone.

To summarize our findings,

- Batters who make good contact tend not to strike out. Low-strikeout players tend to make better contact with both pitches in the zone and pitches outside of the zone.

- Batters who are reluctant to swing at bad pitches (that is, pitches located outside of the zone) tend to walk a lot.

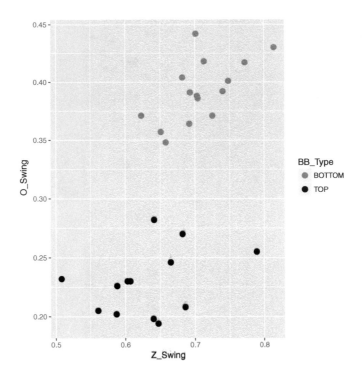

Figure 7.6 Zone and outside swing rates for players with low and high walk rates.

PLATE DISCIPLINE STATISTICS FOR PITCHERS

Two interesting statistics presented in the FanGraphs plate discipline section are the proportion of first pitches that are strikes and the proportion of swinging strikes. These measures are explored from a pitching perspective with the goal of how these relate to the overall quality of a pitcher.

The FanGraphs site lists these statistics for 73 "qualifying" pitchers in the 2016 season who had pitched a sufficient number of innings. Figure 7.7 displays dotplots of the proportions of first-pitch strikes and the proportions of swing and misses for this group.

The proportions of first-pitch strikes average 0.6 and range from 0.55 to 0.70. The six pitchers Michael Pineda, Josh Tomlin, Collin McHugh, John Lackey, Johnny Cueto, and Kyle Hendricks stand out on the high end – these pitchers were especially good in getting a strike on the first pitch in a plate appearance. The bottom panel shows the distribution of proportions of swing and misses. These range from 0.05 to 0.15 with an average of 0.10. Here Noah

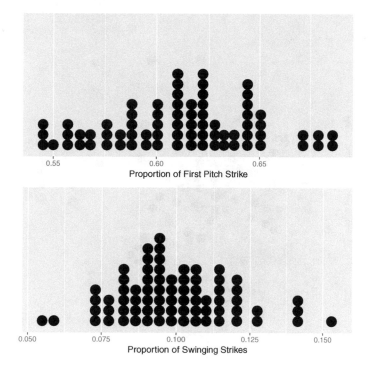

Figure 7.7 Dotplots of proportions of first-pitch strikes and swing and misses among 2016 qualifying pitchers.

Syndergaard, Michael Pineda, Max Scherzer, and Jose Fernandez stand out on the high end – these pitchers are especially able in getting the batter to swing and miss a pitch.

Since both getting a first pitch strike and inducing a swing and miss are both desirable attributes, it is natural to ask how these measures relate to the overall effectiveness of a pitcher. It is more difficult to evaluate the contribution of a pitcher since limiting runs scored is both a function of the pitcher and the defense. However, a general summary of a pitcher's accomplishment is his WAR (wins above replacement) value and so this measure will be used for this study for our summary measure of a pitcher.

Figure 7.8 displays a scatterplot of the proportion of first pitch strikes and proportion of swinging strikes for our pitchers where the size of the plotting point is proportional to the value of the pitcher's WAR statistic. The message from this graph is that both proportions seem useful for explaining WAR – the points of the largest size (largest WAR value) are in the upper-right section of the graph where both proportions are large.

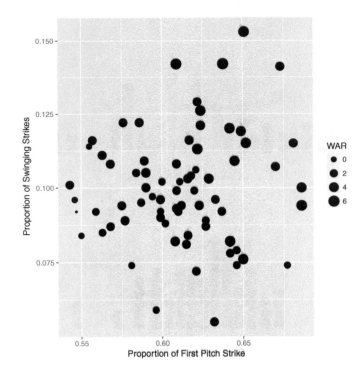

Figure 7.8 Scatterplot of proportions of first-pitch strikes and swinging strikes among 2016 qualifying pitchers where the size of the plotting point is proportional to the WAR statistic.

A regression model is used to explain the variability in the pitcher WAR values based on the two proportion measures. The model has the form

$$WAR = c + a \times First\,Pitch\,Strike + b \times Swinging\,Strike$$

where the constants c, a, and b are chosen so that predicted WAR values are close to the actual WAR observations. Constant values of WAR equal to 0, 2, 4, and 6 from this fitted regression model are shown by lines placed on the scatterplot in Figure 7.9. This graph confirms that both the first strike proportion and the swinging strike proportion are both helpful in predicting WAR. Approximately 40 percent of the total variability in the WAR values can be explained by these two proportion values. For some pitchers, this model provides good predictions. For example, Max Scherzer had a first strike proportion of 0.65 and a swinging strike proportion of 0.15 – this model does well in predicting his season WAR value of 5.6.

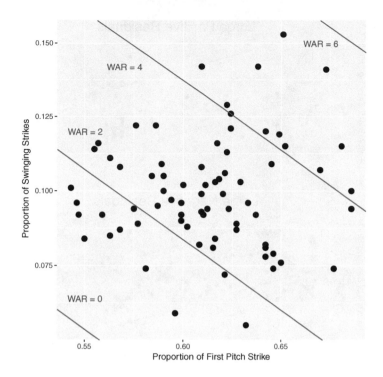

Figure 7.9 Scatterplot of proportions of first-pitch strikes and swinging strikes among 2016 qualifying pitchers. The lines represented predicted values of WAR from the regression model.

However, for some pitchers, this model provides relatively poor predictions. Figure 7.10 redraws the scatterplot – the top panel plots with a "+" the points corresponding to pitchers whose WAR is at least 2 units larger than the prediction, and the bottom panel plots with an "X" the pitchers whose WAR is at least 2 units smaller than the prediction. The deviations of the actual WAR from the predicted WAR values are called residuals.

One of the two points in the top panel (the positive residuals) corresponds to Rick Porcello. Porcello had an above-average first strike proportion, a below-average swing and miss proportion, but a remarkably high WAR value of 5.2 in the 2016 season. Reading an ESPN article, one learns that Porcello changed his pitching style this season. Instead of trying to strike out batters with his four-seam fastball, he used his sinker to induce more ground balls that were turned into outs. Porcello had an excellent 2016 season, finishing with a 22-3 record and winning the Amerlcan League Cy Young award.

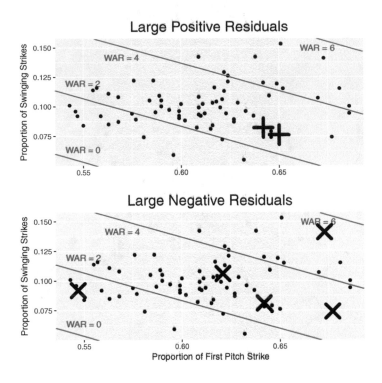

Figure 7.10 Scatterplot of proportions of first-pitch strikes and swinging strikes among 2016 qualifying pitchers. Top panel indicates with a "+" large positive residuals from the fitted model and bottom panel shows large negative residuals with an "X" symbol.

The bottom panel shows points corresponding to large negative residuals – these points correspond to pitchers who had WAR values that were much lower than what would be predicted on the basis of their first pitch strike and swing and miss rates. The Yankees pitcher Michael Pineda had higher first pitch strike and swing and miss rates than Rick Porcello, but a relatively poor season judging by his small WAR value of 3.2. What happened? Reading several blog posts, it appears that Pineda struggled with consistency and often pitched to locations in the middle of the zone. The large residuals for pitchers such as Porcello and Pineda demonstrate that first pitch strikes and swing and miss rates are informative, but they don't capture all of the qualities of pitching that lead to success or failure.

WRAP-UP

The purpose of this chapter is to introduce some modern plate discipline statistics and understand the importance of these statistics in the evaluation of hitting. Batters have to decide whether to swing or not swing at a pitch, and if they swing, they want to make some contact with the ball. Players have varying tendencies to swing, and we have seen that swing rates, especially on balls outside of the zone, can explain batters' tendencies to walk. Also it has been seen that batters are variable with their ability to make contact with a swing, and contact rates are directly related to strikeout rates.

From the pitcher's perspective, it is desirable to obtain a strike and to have the batter swing and miss the pitch. It has been demonstrated that the rates of first pitch strikes and swing and misses are helpful for predicting a pitcher's overall sucess as measured by the WAR statistic. But pitching is a sophisticated endeavor, and it has been seen that pitchers can be ineffective despite having good first pitch strike and swing and miss rates. Likewise, pitchers can be effective despite poor first pitch strike and swing and miss rates. So this study encourages the use of additional measures besides these two plate discipline rates in judging pitcher quality.

CHAPTER 8

Probability and Modeling

INTRODUCTION

One of the enjoyable aspects of baseball is its uncertainty. At the beginning of the season, it is difficult to predict the number of games won for particular teams and the playoffs typically include "weak" teams and exclude "strong" teams. The outcome of a single baseball game is uncertain although the outcome becomes more certain as the game progresses and runs are scored by the two teams. Probability is a way of measuring uncertainty and graphs are used to illustrate various probabilities in this chapter.

The use of probability is first used to address uncertainty in a single game. Assuming the teams are evenly matched, the probability the home team wins at the beginning of the game is 0.5. Every event during the game such as a base hit, strikeout, or fielding play impacts the winning game probability. A special graphical display called a win expectancy graph is used to show how the winning game probability changes after each baseball play. One can graph measures of the game situations called leverages, and this graph illustrates situations that have a large impact on the game outcome. Players have the opportunity to influence the game outcome by good or poor performances during high-leverage situations. Indeed one can measure a player's contribution by how much they added or subtracted to the team's winning probability.

Probability is next used to better understand a season of baseball competition. The 30 teams in MLB have different abilities. These abilities are not observable, but we observe the performances of the teams during the regular season and in the three-tier playoff system. A statistical model can be used to represent baseball competition. By simulating from the model, one can see relationships between team's abilities and their season performances. One can address interesting questions like "how many games will a great team win?" or "what is the probability a great team wins the World Series?"

Fans and media are interested not only in predicting World Series winners, but also predicting individual performances for players. One challenge in this prediction is the so-called "sample-size" effect which says that measures based on a small number of plate appearances exhibit high variation. In particular, batting averages of players from two months of baseball will be variable since the averages are based on a small number of at-bats. But there is information from this two-months data. We use graphs to illustrate how one can predict final season batting averages based on partial-season results.

A BASEBALL GAME

In-Game Win Probabilities

A MLB baseball game is a contest between two teams of relatively equal abilities. When the game starts, the probability that the home team wins is approximately equal to one half. But as the game is played, runners get on base or get out and runs are scored, and the probability of the home team will change. The *win expectancy* is the probability of the home team winning the game as a function of the current score, inning, runners on base, and number of outs. In Chapter 3, we explored graphically the notion of runs expectancy defined by the potential to score runs in an inning given the current runners on base and the number of outs. The values of the runs expectancy are obtained by use of historical baseball data. In a similar fashion, by use of historical data, one can estimate the probability a team will win the game using the runners/outs situation together with information about the inning and the current score. When the win expectancies are graphed against the play number, one obtains the well-known *win expectancy graph* used by the Baseball Reference and FanGraphs web sites to summarize the "important" events in a baseball game.

A win expectancy graph will be illustrated for a famous game, Game 7 of the 2016 World Series between the Chicago Cubs and the Cleveland Indians where the Cubs won 8 to 7. There were 96 plays (plate appearances) in this game and Figure 8.1 constructs a graph of the probability that the home team (the Indians) wins the game after each plate appearance.

At the beginning of the game, it is assumed that the probability of the Indians winning the game is 0.5. The first plate appearance

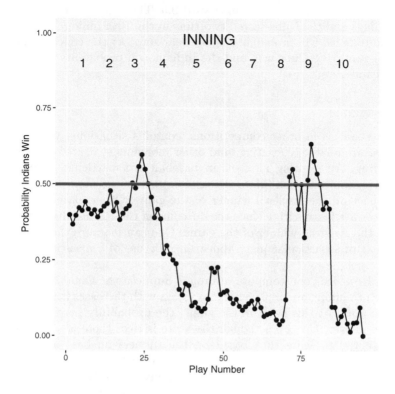

Figure 8.1   Win probability graph for Game 7 of the 2016 World Series between the Cleveland Indians and the Chicago Cubs.

in the 1st inning Dexter Fowler, the Cubs leadoff hitter, homered, giving the Cubs a 1-0 lead. The probability of the Indians winning the game (based on historical data) has now changed to 0.398. One can measure the contribution of this play toward the goal of winning the game by the change probability. Fowler's home run decreased the Indians win probability from 0.50 to 0.398 and so he contributed $0.50 - 0.398 = 0.102$ toward the probability of the Cubs winning.

The story of Game 7 of the 2016 World Series can be explained using the win expectancy graph of Figure 8.1. In the first few innings, the Cubs had the lead and the Indians' win probability stayed below 0.5. But the Indians scored a run in the 3rd inning and the Indians' win probability briefly rose over 0.5. But then the Cubs scored runs in the 4th, 5th, and 6th innings and the Indians' win probability drifted toward 0. The Indians had a stirring three-run comeback in the 8th inning, tying the game, and the win probability moved again toward 0.5. The game went to extra-innings and the Cubs scored two runs in the 10th inning, lowering the Indians' win probability again near zero. At the conclusion of the game, the Cubs won and the Indians win probability actually became zero.

Leverages

Baseball, as in other competitions, contains situational moments that are relatively exciting, and other moments that are relatively boring. An exciting moment in baseball is a particular pitcher-batter confrontation (plate appearance) whose outcome has a large impact on the eventual winner of the game. Other plate appearances are less exciting since the outcome of the PA has little effect on the eventual winner of the game. One can measure the importance of a particular plate appearance by use of a measure called *leverage*.

How does one compute leverage? Suppose the game is tied in the 6th inning and a player comes to bat with the bases loaded and two outs. At this point in the game, the probability his team will win is over 0.5 – even though the score is tied, there is a positive potential to score runs based on the runners on base and outs situation. What can happen in this situation? The batter can get a base hit, scoring runs, and contributing a positive value to his team's win probability. Or the batter can get out, not scoring runs, and contributing a negative value to his team's win probability. The leverage is the weighted average of the sizes of all of the possible changes in win probability, where the weights are proportional to the chances of these events happening. In this situation with a tie game and runners on base, the leverage value would be relatively large, since there are events that can cause large changes in the win probabilities. In contrast, a situation where the bases are empty with two outs would be a case of low leverage. The different possible outcomes such as a base hit and an out would have modest changes in the team's win probability.

One can understand the exciting moments in a game by graphing the leverage values as a function of the play number. Figure 8.2 displays the leverages of the 96 plate appearances in Game 7 of the 2016 World Series. What do we see in this graph? The leverage values are relatively small in the early innings and there is a cluster of higher values in innings 3 and 4, indicating there were exciting moments with men on base in those innings. After the Cubs grabbed the lead, the leverage values were small in innings 5 through 8, indicating that the Indians have few opportunities to change the win probability in those innings. After the Indians scored runs to

tie the game in the bottom of the eighth, the leverage values increased dramatically which makes sense for a tied game where the
final outcome could be determined by a single swing of the bat.

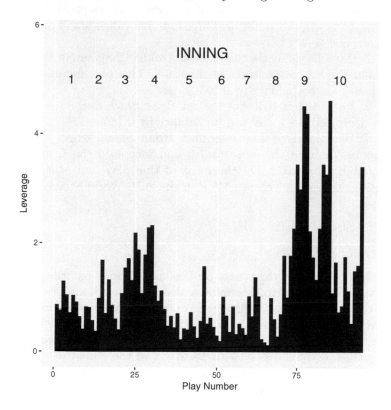

Figure 8.2 Leverage plot for Game 7 of the 2016 World Series.
The large values correspond to the plate appearances during
the game where the outcome of the plate appearance could
have a large change in the win probability.

Win Probability Contributions

Leverages are helpful for understanding the potential for scoring
runs that can impact the game's win probability. But it does not
measure the game plays that actually had a large impact on a
team's win probability. One can measure the importance of game
plays by use of the change in win probabilities. By graphing these
win probability contributions, one can visualize the important individual accomplishments in the game. This display is called a win
probability added graph.

Figure 8.3 displays the changes in win probabilities for Game 7 of the 2016 World Series as a function of the play number. Positive values correspond to plays that increased the batting team's win probabilities and negative values correspond to plays that decreased the win probabilities. Three extreme values stand out that correspond to key plays in this game.

1. Rajai Davis' home run against Aroldis Chapman in the bottom of the 8th inning tied the game 6-6. This play increased the Indians' win probability by 0.405.

2. Javier Biaz's strikeout against Brian Shaw in the top of the 9th decreased the Cubs' win probability by 0.189.

3. Ben Zobrist's double against Brian Shaw, scoring Albert Almora, changed the score to 7-6 and increased the Cubs' win probability by 0.322. On the basis of this play and other good hitting, Zobrist was chosen MVP of the 2016 World Series.

A BASEBALL SEASON

Playoff Structures

Baseball is one of the oldest professional sports, but the design of the competition to determine a season's "best team" has gone through a number of changes over the years. To describe the changes, let's compare the competition format in the 1968 season with the current 2017 season. In the 1968 season, there were 20 teams, divided evenly with 10 teams in the American League and 20 in the National League. All teams played a schedule of 162 games where each team played teams in its own league. The team with the most wins in its league became the League Champion and the National and American League Champions faced off in a best of seven World Series. (In passing, in 1968, the Detroit Tigers defeated the St. Louis Cardinals 4 games to 3 to become the 1968 World Series champion.)

Currently, in the 2017 season, there is a very different format to decide the "best team" in baseball. There are currently 30 teams, 15 in each league that are divided into three divisions of five teams. For example, the American League is divided into the East, Central, and West divisions. At the end of the season, five teams in each league advance to the playoffs including the three division-winners and two "wild card" teams with the best records among all of the teams not winning a division. At this point, there are four levels of playoffs:

1. In each league, the two wild-card teams play in a single-game Wild Card playoff.

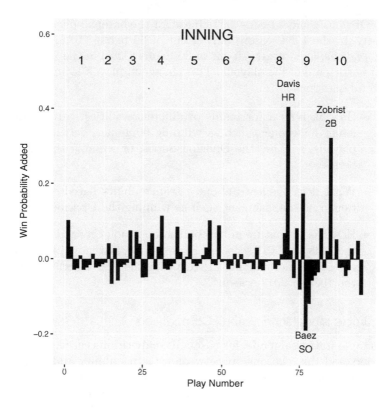

Figure 8.3 Win probability added graph for Game 7 of the 2016 World Series. The value plotted represents the addition to the win probability for the team that was batting. Three extreme values during this game are labeled.

2. In each league, two of the division-winners play a "best of five" game playoff. In addition, the other division winner plays the winner of the Wild Card game in a "best of five" series. These playoffs are called Division Series, abbreviated as ALDS1 and ALDS2 (American League), and NLDS1 and NLDS2 (National League).

3. In each league, the winners of the Division Series play off in a "best of seven" league championship. These are called ALCS and NLCS for American League Championship Series and National League Championships, respectively.

4. The two winners of the Championship Series play off in a best-of-seven World Series.

Obviously, more teams participate for the baseball playoffs currently. In the 1968 season, only 2 out of 20 teams (10%) were in the playoffs; in contrast, 10 out of 30 teams (33%) in the 2017 season participate in the playoffs. This change in playoff format raises several questions:

- How likely is it for teams of different abilities to reach different milestones, such as winning 90 games, getting in the playoffs, winning the Division Series, or winning the World Series?

- What does one learn about a team's ability based on a particular accomplishment such as winning the Division Series?

- How has the nature of baseball competition changed between the 1968 and 2017 seasons? For example, is it more likely that one of the most talented teams wins the World Series during the 1968 or 2017 seasons?

A Simple Model for Baseball Competition

A simple statistical model for baseball competition is used to better understand the relationship between a team's ability and its performance during a baseball season. There are currently 30 baseball teams. Suppose each team is assigned a talent denoted by T – a talent is a single number that describes the overall strength of a team including pitching, batting, and defense. Since there are 30 teams, we have 30 talents that we denote by $T_1, ..., T_{30}$. It is assumed that these talents follow a bell-shape (normal) curve with mean 0 and standard deviation 0.20. (The value of the standard deviation is chosen so that the predicted number of winning and losing games for the teams resemble the actual winning records.)

Figure 8.4 displays the normal curve that represents the distribution of baseball talents. The vertical lines are used to divide the team talents into five categories. The "poor" teams represent the bottom 10% of the talent distribution. The "below average" teams are between the 10th and 30th percentiles, the "average" teams fall between the 30th and 70th percentiles, the "above-average" teams are between the 70th and 90th percentiles, and the "great" teams are in the top 10% of the talent distribution.

Given the talents of two baseball teams, say team A and team B, the Bradley-Terry model gives a simple formula for the probability team A defeats team B in a single game – it is given by

$$P(Team\ A\ wins) = \frac{\exp(T_A)}{\exp(T_A) + \exp(T_B)}.$$

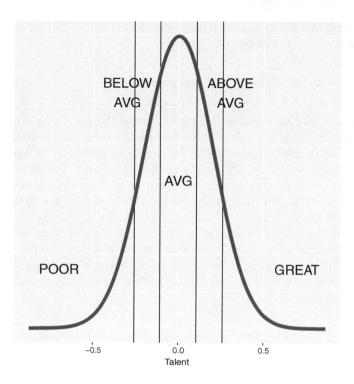

Figure 8.4 Talents of MLB teams. The vertical lines are used to divide the teams into "poor", "below average", "average", "above average", and "great" categories.

Simulating a Baseball Season

This Bradley–Terry model gives us a simple way for simulating an entire 2017 baseball season. First, one simulates the talents of the 30 teams. Then using the actual published 2017 baseball schedule, one simulates the outcome of each game using the Bradley–Terry model with the given team strengths. This is a pretty simplistic model. It does not distinguish a team's batting ability from its pitching ability, and it also assumes that a team's strength does not change over the course of the six-month season. But this model approximately matches the distribution of games won by 30 teams over a season of 162 games. For that reason, this can be viewed as a reasonable model for baseball competition.

This method is used to simulate 5000 baseball seasons. By repeating this simulation a large number of times, one can explore the relationship between a team's ability T with the number of games the team wins in a season. Also, one can explore the relationship between a team's ability with its success or lack of success in advancing in the playoffs.

Performance of Teams of Different Abilities

The first goal of a team is to win as many games during the 162-game season as possible to give the team the greatest chance of getting into the playoffs. Figure 8.5 constructs density plots of the number of season games won for teams in each one of the five talent levels. We see, for example, that the "poor" teams tend to win between 60 and 70 games – there is a small chance that the team will win over half of its games. (Vertical lines are drawn at 81 that reflect a team that wins exactly half of its games.) A team of "average" ability will tend to win between 75 and 85 games, and a "great" team will tend to win between 90 to 100 games. Also it is interesting to note that the "below average", "average", and "above-average" teams are similar with respect to the number of games won.

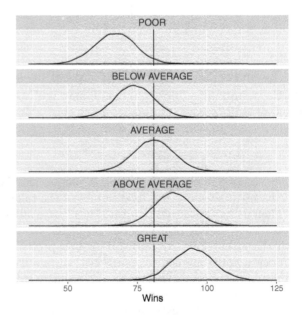

Figure 8.5 Density plots of the number of season wins of teams of different talent levels from the simulation. The vertical lines are drawn at 81 which reflects a team that wins exactly half of its games during the season.

Although a good performance in the 162-game season will get a team in the playoffs, it is interesting to see how teams of different abilities perform in the different levels of playoffs. Figure 8.6 shows the probabilities that teams of different talent groups reach each of the following milestones:

- "GET IN": get into the first level of the playoffs either as a division winner or a wild card team

- "WIN LDS": win the first "best-of-five" league division series

- "WIN LCS": win the league championship series

- "WIN WS": win the World Series and be crowned the MLB champion

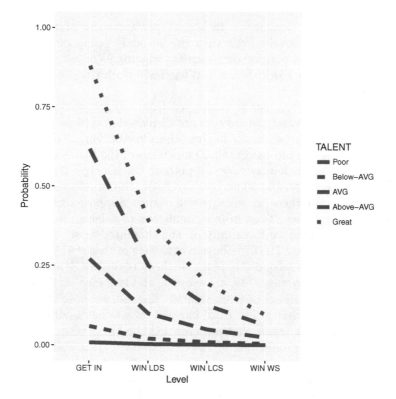

Figure 8.6 Probabilities that teams of different talent levels reach different milestones.

There are several interesting insights from Figure 8.6. The ordering of the curves is what one would expect – a "great" team is more likely to do well in achieving a particular playoff mile-stone, followed by "above average", "average", "below average", and "poor" teams. For example, the probability a great team reaches the playoffs is over 0.85 contrasted with an average team which has only a 0.25 probability of getting to that milestone. The probability of getting to the next milestone dramatically drops for a team of a particular ability level. For example, although it is very likely that a great team makes the playoffs, the probability the team wins the LDS is only about 0.37, and this great team only has a 0.12 probability of winning the World Series. These large probability drop-offs are a by-product of the short best-of-five and best-of-seven game series that are currently used in the MLB playoff system.

Learning About Team Abilities from Their Performance

In the previous section we looked at performances of teams of different ability levels. Let's turn this around – suppose you observe a particular performance such as "winning 95 or more games" or "winning the World Series"? What have we learned about the team's ability?

These questions can be addressed from our simulation of the baseball seasons and the answers are displayed graphically in Figure 8.7. Before the season begins, the distribution of talents is displayed by the top bar graph. Ten percent of the teams are poor, 20 percent are below average, 40 percent are average, 20 percent are above average, and 10 percent are great. But after a team completes a season, these percentages will change. Suppose a team wins 95 or more games. Then in our simulation of talents and performances of teams, we focus only on the simulated teams that win 95 or more games. The distribution of talents of these "95+" teams is shown as the second bar plot. Of these teams, 54% are great, 34% are above average, 12% are average, and it is rare for the team to have below average or poor abilities. Instead, suppose the team wins the World Series. The third bar plot shows the distribution of abilities of these teams that are crowned the "best team in base-ball." Interesting, only 29% of these teams have great ability, 38% are above average, 30% are average, and 3% are below average. There are several interesting take-aways from this analysis. First, it is more likely for the team to be great if they win 95 or more games than if they win the World Series. Second, a high proportion of "not great" teams win the World Series. This conclusion is likely a by-product of the many-tier playoff structure in modern baseball.

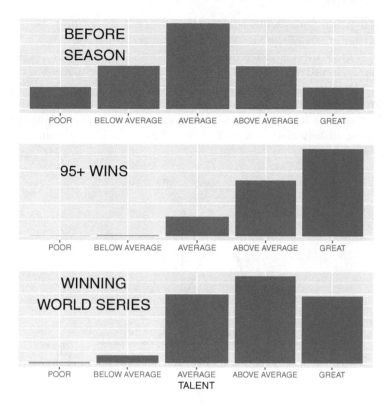

Figure 8.7 Talent levels of baseball teams before the season, after winning 95 or more games, and winning the World Series.

Comparison of 1968 and 2017 Playoff Systems

To expand the last statement, suppose we also perform many simulations using the 1968 schedule and 1968 playoff schedule and the same talent distribution of teams. Suppose two groups are compared in the many simulations – one group consists of all teams that won the World Series and the second group consists of all teams that did not win the World Series. Figure 8.8 presents density graphs of the talents of the two groups of teams for each of the 1968 and 2016 seasons. The solid line is the normal curve with mean 0 and standard deviation 0.2 that describes the overall distribution of talents of the teams. What is interesting is the talent curves for the teams that won the World Series. The vertical line corresponds to the division line between above average and great teams. In 1968, this line falls approximately in the middle of the density curve – the conclusion is that half of the World Series winners in 1968 were great teams. In contrast, it is less likely for a

2016 World Series winner to be great – in our work above, we saw that only 29% of the WS winners had great talent.

Why is there a difference between the 1968 and 2017 proportions? Recall that the 1968 season had a shorter playoff structure (two levels of best-of-seven playoffs), contrasted to the wildcard/division series/league/World Series current playoff structure. Currently more teams get into the playoffs, but it also means that it is less likely that a World Series winner AVG is a team with great talent.

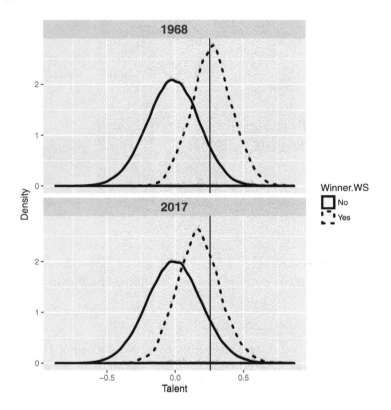

Figure 8.8 Density graphs of talents of World Series winners and non-WS winners for the 1968 and 2017 seasons. The vertical line represents the breakpoint between teams of above average and great talents.

BATTING AVERAGES

Batting Averages After Each Month

The baseball season is scheduled from early April to early October, a period of six months. As the season progresses, there are tables of leaders of various batting and pitching measures. It can be difficult to make sense of these leaderboards early in the season, since these measures are based on a relatively small number of opportunities. Here this general issue is illustrated using the traditional measure of batting performance, the batting average.

A batting lineup consists of nine players and since there are 30 baseball teams, there are approximately 250 hitters who play regularly and get many opportunities to bat. A batting average is defined by $AVG = H/AB$ where H is the number of hits, and AB is the number of at-bats. Figure 8.9 constructs a dotplot of the batting averages for the 250 regular players after each month of the 2016 season.

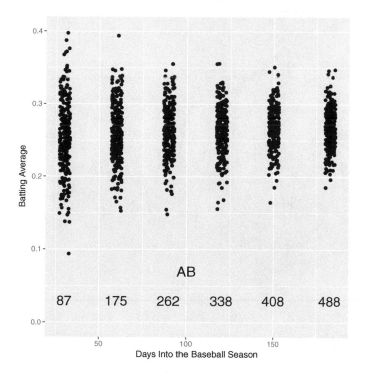

Figure 8.9 Scatterplot of batting averages of 250 players after each month of the 2016 season. The bottom labels give a typical number of at-bats for each month.

The vertical strip on the left is a jittered dotplot of the AVGs after 30 days, the next vertical strip is a graph of the AVGs after 60 days, and the last vertical strip is a display of the AVGs at the end of the season. We display a representative number of at-bats at the bottom of each vertical strip. So the "30-day" AVGs are batting averages with roughly 87 at-bats, the "60-day" AVGs are batting averages with 176 at-bats, and so on. The main take-away from this graph is that there is more spread in the batting averages early in the season, and the spread of the averages decreases as one progresses through the season.

To make this pattern of these monthly batting averages more clear, Figure 8.10 displays parallel violin plots of the AVGs after each month of the season. The violin plot is better than the jittered dotplot in showing the concentration of the AVGs – a greater concentration is reflected by a wider section of the violin. Figure 8.5 shows the wide spread of the 30-day batting averages. In contrast, at the end of the season, most of the averages are concentrated about the middle value of 0.260.

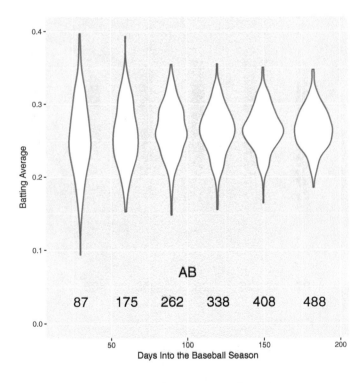

Figure 8.10 Violin plots of batting averages of 250 players after each month of the 2016 season.

Let's focus on the 30-day batting averages. Figure 8.11 displays a histogram of the batting averages and overlays a normal (bell-shape) curve with mean 0.254 and standard deviation 0.05 – this curve appears to be a pretty good match to the distribution of batting averages.

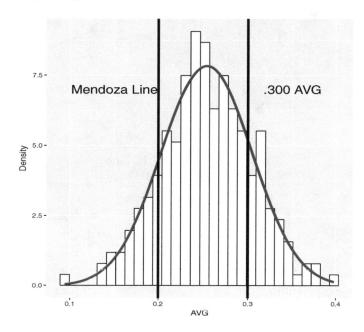

Figure 8.11 Histogram and normal curve fit to the batting averages of MLB regulars after 30 days of the 2016 season.

The vertical lines in Figure 8.11 indicate special batting averages. The line at AVG = 0.200 is the Mendoza line, deriving from Mario Mendoza who was a pretty weak hitting shortstop in the 1970s. This line is an indication of poor hitting and it is rare in modern baseball for a hitter to hit below the Mendoza line for a full season. In contrast, the second line at AVG = 0.300 is a mark for good hitting – most major leaguers would be pleased to have a batting average over 0.300 at the end of the season. Figure 8.6 indicates that a good number of batters after 30 days have "below Mendoza line" averages and averages of 0.300 or higher. Looking more carefully at the data, we see that five hitters actually had batting averages under 0.150 at 30 days and Logan Morrison had a 6 for 64 = 0.094 average. At the high end, there were five "30-day" averages of 0.350 or higher and Martin Prado was 33 for 83 for a robust 0.398 average.

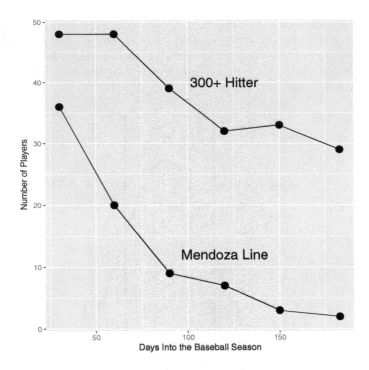

Figure 8.12 Number of hitters below the Mendoza line and above 0.300 at different times during the 2016 season.

Although we see some interesting small and large batting averages early in the season, these interesting AVGs become less common later in the season. To make this clear, I collected

• The number of AVGs below the Mendoza line (below 0.200)

• The number of 0.300 AVGs or higher

among the 250 players at each point in the season. Figure 8.12 displays the number of hitters in the low and high categories against the number of days into the baseball season. Note that the number of Mendoza hitters drops off sharply and there were very few hitters with AVG under 0.200 at the end of the season. Looking back at the data, there were only two hitters– Derek Norris (0.186) and Ryan Howard (0.196)– in this group. Also, the number of 0.300+ hitters decreased over the season. There were 45 hitters in this group after 30 days and only 28 300+ hitters at the end of the 2016 season. By the way, DJ LeMahieu had the highest AVG of 0.348 in 2016.

Predicting Final Averages

On June 1, 2016, Xander Bogaerts had 77 hits in 222 at-bats for a 0.347 batting average. What is a reasonable prediction of Bogaerts' batting average for the entire 2016 season? This section describes a prediction method by use of several graphs.

To start, what is the relationship between a player's two-month batting average (the average on June 1) and his final season batting average? Figure 8.13 constructs a scatterplot of the two-month AVG and the season AVG for all players who had at least 150 at-bats on June 1. One sees that there is a strong positive relationship. This makes sense – players with high batting averages in the first two months would be expected to have high season averages.

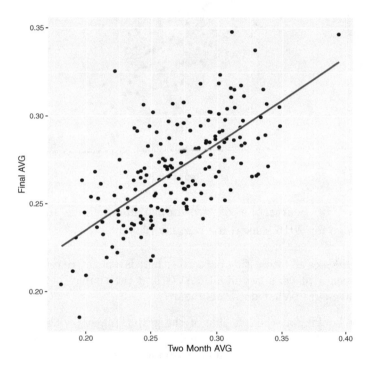

Figure 8.13 Scatterplot of two-month and final batting averages in 2016 season for regular players.

But how will a player's season average *differ* from his two-month average? One can address this question by graphing the change in batting average

$$CHANGE = AVG_{season} - AVG_{2-month}$$

against the two-month batting average for the same players in Figure 8.14.

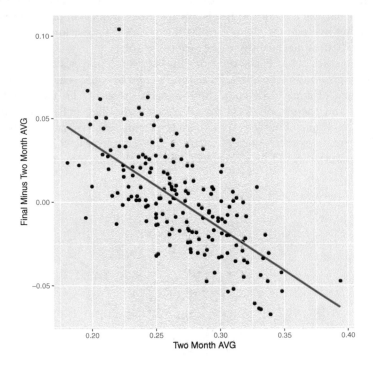

Figure 8.14 Scatterplot of two-month and change in batting average for 2016 season for regular players.

One sees an interesting pattern – there is a negative relationship between a player's two-month AVG and the change between the two averages. What does this mean?

- Looking at the left side of the graph, players with low two-month batting averages will tend to have positive change, that is, have higher final season averages.

- On the right side of the graph, we see players with high two-month AVGs will tend to have negative change, that is, have lower final AVGs.

Another way of saying this is that a player's two-month batting average will tend to move toward the average. Players with high two-month AVGs will tend to decline toward the end of the season, and players with low two-month AVGs will tend to improve at the end of the season. In statistical jargon, this phenomenon is called the *regression effect* or *regression to the mean*.

How can one find a good prediction of a player's final batting average knowing his AVG after two months? Since we want to adjust the two-month average toward the overall batting average, a reasonable type of prediction has the form

$$Prediction = F \times AVG_{2-month} + (1 - F) \times AVG_{all},$$

where AVG_{all} is the mean batting average for all players in the first two months, and F is a fraction between 0 and 1 that indicates how much you move the two-month average toward the overall average.

How does one find the best value of the fraction F? For a specific choice of F, one can use the formula to predict the season batting average for our group of hitters and see how close the predictions are to the actual 2016 season AVGs. One measure of closeness, the prediction error, is the sum of squared errors:

$$Prediction\,Error = \sum (Prediction - AVG_{season})^2.$$

We tried many values of F between 0 and 1 and Figure 8.15 graphs the prediction error against the fraction F. One wants to find the fraction value that makes the prediction error as small as possible. The best choice of F turns out to be 0.49. The overall batting average for the two months of the 2016 season is 0.270, so the best prediction of a player's final season AVG is

$$Prediction = 0.49 \times AVG_{2-month} + 0.51 \times 0.270.$$

Recall Xander Bogaerts had a 0.347 average after two months – one predicts his final AVG to be

$$0.49 \times 0.347 + (1 - 0.49) \times 0.270 = 0.308.$$

Essentially one is shrinking or adjusting Bogaert's two-month average about 50% of the way toward the overall batting average. (In passing, it is interesting to note that Bogaert finished the 2016 season with a 0.294 AVG.)

Figure 8.16 displays the two-month batting average and the predicted final season average for 10 randomly selected players. This graph shows how one should adjust a player's midseason average to get good predictions at his season batting average. This does not mean that the prediction for one particular player will be closer than his two-month average to his season AVG. But across all of the players, the overall accuracy of all the predictions will be better using this "shrinkage to the mean" method.

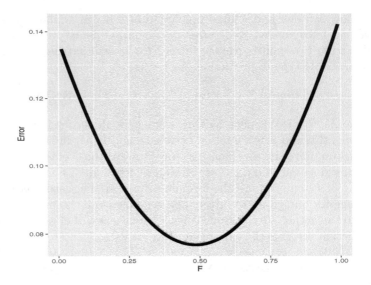

Figure 8.15 Plot of prediction error of the batting average estimate for different values of the fraction F. One can see that the smallest value of the prediction error is attained about $F = 0.5$.

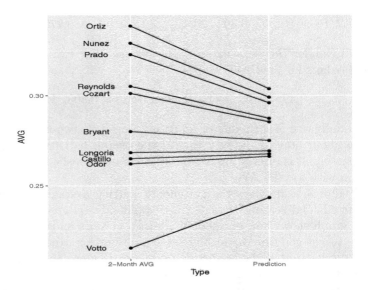

Figure 8.16 Two-month batting averages and prediction using the shrinkage estimate for 10 randomly selected players.

The amount of shrinkage (value of F) depends on the number of at-bats in the first part of the season. For example, if you were using AVGs from only one month of the season (instead of two months), then the value for the best F would be smaller than 0.49 – the prediction would shrink the one-month AVG further toward the overall batting average.

WRAP-UP

Baseball is fun to watch since much of the action is unexpected. Even though a team may be losing for many innings, it is possible for a team to come back at the end so fans will stay until the last pitch in the ninth inning. A win expectancy graph is a helpful way to visualize the ups and downs in a team's fortune during the game and to document the game moments that had a large impact on the final outcome. These graphs also can be helpful from the coaching perspective. A coach may make a decision such as attempting to steal a base by thinking about the probabilities of winning the game if the stolen base attempt is successful or fails.

Baseball is one of the more competitive of professional team sports and the model simulation described in this chapter helps one to better understand the characteristics of this competition. It is a fact that chance variability has much to do with the results of baseball games and even teams of average ability have a chance of making it into the baseball playoffs. The role of randomness becomes even more prominent during the playoff since there is much uncertainty in the outcome of a seven-game playoff series. As many writers note, baseball playoffs are essentially a crap shoot.

Prediction of individual player performances is difficult although teams devote much of their energy in drafting and signing players that they think will be successful in the Major Leagues. Graphs are used here in demonstrating the so-called regression effect – players tend to regress or move toward the average. But one can apply this knowledge to develop reasonable predictions of final season statistics based on statistics from the first couple of months of the season. Good predictions of future performance are important, especially when one is using historical data to make decisions about signing free agents during the winter months.

CHAPTER 9

Streakiness and Clutch Play

INTRODUCTION

The media keeps track of a range of measures for batters and pitchers, but they also keep track of specific hitting and pitching patterns call streaks and slumps. Players are considered "hot" if they get hits in a large number of consecutive games. On the other hand, teams are concerned when a player is hitless in a large number of consecutive at-bats, the so-called "ofer" statistics. (These slumps are reported as "0 for 20" or "0 for 15" hence the phrase "ofer.") This chapter illustrates the use of graphs to visualize patterns of hot and cold hitting of players. One challenge in interpreting these graphs is that random data, analogous to sequences of flips of a fair coin, can appear streaky. A special graph is illustrated that can help distinguish between coin-flipping data and streaky hitting data from ballplayers.

Players are often characterized as streaky hitters. Similarly, players can be described as clutch as they appear to perform especially well during important situations during the game. We first explore different baseball situations (runners on base and outs) and use the concept of leverage to study the importance of different situations. One can estimate the number of runs a player produces if his plate appearance outcomes are not related to the game situations. By comparing the actual number of runs produced with this estimate, one can measure which players are especially good in getting hits in important situations.

STREAKY HITTING

Rug Plots

A popular measure of batting performance is the batting average $AVG = H/AB$. During the 2016 season, Neil Walker and Nori Aoki had similar AVG seasons – Walker had 116 hits in 412 at-bats for a $116/412 = 0.282$ average, and Aoki had 118 hits in 417 at-bats for a $118/417 = 0.283$ average.

Although Walker and Aoki had essentially the same 2016 batting average, the patterns of their hitting throughout the season actually were very different. We can learn about these patterns by recording the at-bat numbers of the base hits for both players. Walker came to bat 412 times and we number the at-bats 1 through 412. Walker's at-bat numbers are recorded where he got a base hit – he started the season by getting 10 hits in at-bat numbers 6, 8, 11, 12, 24, 27, 30, 32, 34, and 40. Figure 9.1 displays "rug plots" of the base hits of Walker and Aoki where the at-bat numbers of base hits are plotted by vertical lines.

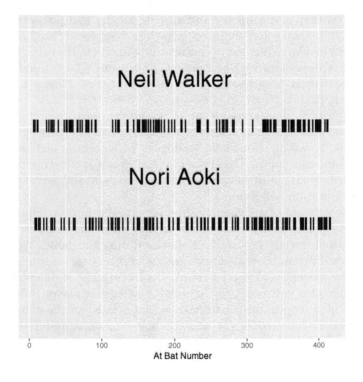

Figure 9.1 Rug plots of the hit occurrences for two players in the 2016 season.

This figure illustrates different patterns of hitting for two players with similar season batting averages. Walker's plot shows notable gaps – at approximate at-bat numbers 100, 230, and 300 there are white spaces indicating it was hard to get a base hit during those periods. In contrast, Aoki appears to have a more consistent pattern of hitting throughout the season. The at-bat hit numbers seem to be evenly distributed throughout the season with only one noticable white space about at-bat number 70. The take-away message from Figure 9.1 is that Walker had more hitting slumps and streaks than Aoki during the 2016 season.

Moving Average Plots

One can learn more about the pattern of hitting of a specific player by computing batting averages over short time intervals. For example, let's revisit Neil Walker's hitting during the 2016 season. If one looks at Walker's performance from the 40th to the 89th at-bats, he had 18 hits for a $18/50 = 0.360$ batting average. In contrast, if one looks at Walker's 100th to 149th at-bats, he had only 9 hits for a $9/50 = 0.180$ average. Suppose one computes Walker's batting average for all periods of 50 at-bats; so one computes his average for at-bats 1 through 50, for at-bats 2 through 51, for at-bats 3 through 52, and so on. If one plots these "moving averages" against the average at-bat number, one obtains the moving average display in the top panel of Figure 9.2. The horizontal line in the middle of the plot represents Walker's season AVG of 0.282. Shaded areas above the horizontal line represent periods where Walker was "hot" and shaded areas below the line represent 50 at-bat periods where Walker was "cold". From this graph, we see that Walker started hot, had a cold period, a shorter hot period, a long cold period, and a strong hitting period at the end of the season.

Let's contrast the moving average plots of the two players – in each plot we are using 50 at-bat periods. Aoki also displays periods of cold and hot hitting; for example, he was relatively cold at the beginning of the season and was generally hot at the end of the season. But the sizes of the cold and hot areas are much smaller for Aoki than for Walker. These moving average displays confirm that Walker's pattern of hits and outs during the season was much more volatile than Aoki's pattern of hitting. Another way of saying this is that Aoki was more consistent in his short-term hitting than Walker.

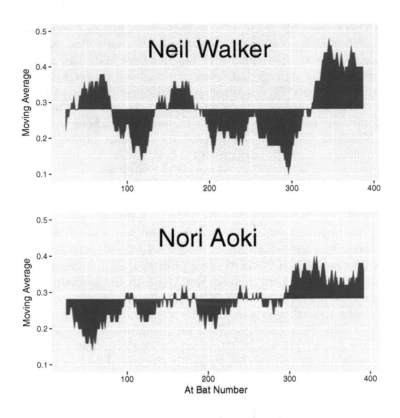

Figure 9.2 Moving average plots of the batting averages using a window of 50 at-bats for two players in the 2016 season.

Looking at Spacings

Recall that Neil Walker started the 2016 season by getting hits in at-bat numbers 6, 8, 11, 12, 24, 27, 30, 32, 34, and 40. Another way to view Walker's hot and cold hitting is to look at the spacings, the number of outs between consecutive hits. Walker's first hit was on the 6th at-bat, so the number of outs before the first hit was 5 – the first spacings value is 5. Then there was one out before Walker's next hit (on the 8th at-bat) – the next spacings value is 1. Walker's next hit was on the 11th at-bat; there were two outs between the 8th and 11th at-bats so the next spacings value is 2. Looking at the at-bat numbers of the first 10 hits

$$6, 8, 11, 12, 24, 27, 30, 32, 34, 40$$

we see that the values of the first 10 spacings are

$$5, 1, 2, 0, 11, 2, 2, 1, 1, 5.$$

All of the spacings values were collected for Neil Walker and Nori Aoki for the 2016 season. Figure 9.3 displays parallel dotplots of the spacings for the two hitters. Generally the shapes of the distributions of the spacings are similar for both hitters – it is more common to have small spacings than large spacings. But under more careful examination, one notices some differences. Walker was more likely to have a spacing value of zero, and also Walker appears to have more extreme spacings values than Aoki.

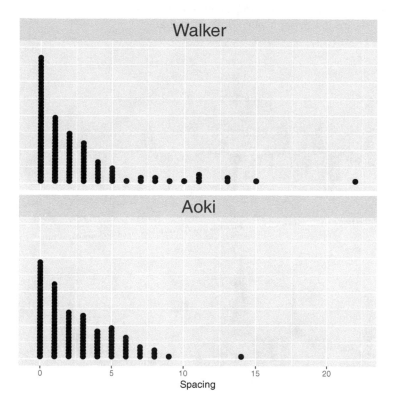

Figure 9.3 Dotplots of the spacings between successive hits for two players in the 2016 season.

Random Data

One difficulty in interpreting patterns in hitting data is that random data can look unusually streaky even when there are actually no "real" hot or cold tendencies among players. This idea can be illustrated by considering random hitting data from hypothetical players who are truly consistent.

Suppose a player is a truly consistent hitter in the sense that the probability of getting a hit on a single at-bat is 0.300. The chance of a base hit on any at-bat is always 30% and the chance of getting a hit on, say, the 200th at-bat is not dependent on the player's success or lack of success in the previous at-bats.

Suppose we have four of these consistent 0.300 hitters and each has 500 at-bats during a season. What type of streaky patterns do we see in their hitting? One can simulate hitting data for these players and display moving average plots of their day-to-day hitting in Figure 9.4 using a period of 50 at-bats.

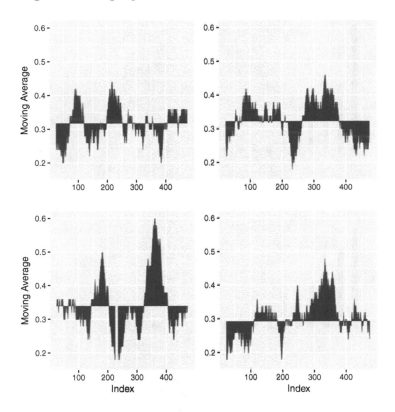

Figure 9.4 Moving average plots of batting average using a window of 50 at-bats for four simulated consistent hitters.

These players are the ultimate consistent hitters since they have the same probability of a hit on every at-bat. But Figure 9.4 demonstrates that these players can display some interesting streaky behavior. For example, simulated player 2 (upper right panel of Figure 9.4) had a 50 at-bat batting average as small as 0.200 and as large as 0.450 during the season. Simulated player 3 (lower left panel) actualy had a batting average as high as 0.600 during 50 at-bats. Although their hitting abilities are consistent, they certainly don't perform consistently through the season.

A Geometric Plot

By graphing the moving averages of hitting data from truly consistent hitters, one sees that these graphs don't appear to be that helpful in detecting unusual streakiness. But an alternative graphical display on the spacings, that is, the gaps between consecutive hits, is potentially useful in the study of streakiness.

Let S denote the spacing between consecutive at-bats. If the hitter is truly consistent with a constant probability p of a hit, then S has a geometric density where the probabilities are given by the formula

$$PROB = (1 - p)p^S, S = 0, 1, 2, ...$$

Taking natural logarithms of both sides of the equation, we get the equality

$$\log PROB = log(1 - p) + S \times \log(p), S = 0, 1, 2, ...$$

Note that the logarithm of the geometric probability is a linear function of the spacing value S where the slope is $\log p$. This motivates the following "geometric plot". Given a large number of spacing values, construct a frequency table where you find the frequency for each possible spacing value. Draw a scatterplot of the spacing values against the logarithms of the corresponding frequencies. If the data comes from a geometric distribution, the points should fall along a straight line.

Figure 9.5 illustrates the use of geometric plots for the simulated consistent data used in the earlier investigation. For each player, the spacings were computed and tabulated, and the points represent the spacing values plotted against the logarithm of the frequencies. A smoothing curve is used to show the point pattern and is contrasted with a "best" straight line fit through the data. Since the smoothing curves are close to straight in each example, the message from the figure is that the points tend to follow a straight line for simulated "true" geometric data. From experience from looking at many plots, a curvature pattern in a geometric plot indicates streaky patterns in hitting data that do not follow a geometric distribution.

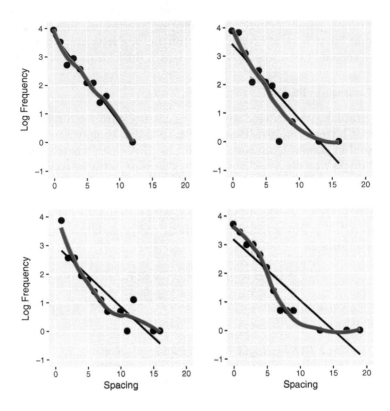

Figure 9.5 Geometric plots of the spacings between hits for four simulated consistent hitters. A smoothing curve is used to show the basic pattern of the points and the line represents a best straight fit through the data. The points generally follow along a line for spacings where there is a constant probability of success.

Revisiting the Two 2016 Hitters

We return to our two hitters Neil Walker and Nori Aoki who had similar batting averages in the 2016 season. Figure 9.6 displays geometric plots of the spacings (outs between consecutive base hits) for these two players. Note that Aoki's points appear to follow a line and Walker, in contrast, has points that seem to deviate far from the line. The message is that the streakiness in Walker's data does not follow the patterns if he was truly a consistent hitter.

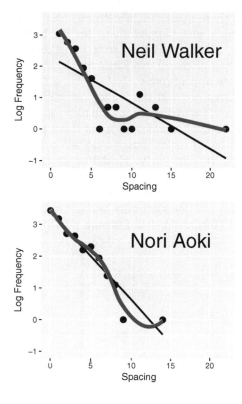

Figure 9.6 Geometric plots of the spacings between hits for two players in the 2016 season. The points in Walker's plot don't follow a line indicating his spacing values don't follow a geometric distribution.

No-Strikeout Streaks

During the 2017 season, Mookie Betts of the Boston Red Sox had a remarkable accomplishment – he had a streak of 129 consecutive plate appearances, spread over the 2016 and 2017 seasons, where he did not have a strikeout. This is one of the longest streaks of this type in modern baseball history. It might be more remarkable since we saw in Chapter 1 that strikeout rates in Major League Baseball have been rising in recent seasons.

This accomplishment raises several questions:

- What are typical "long" strikeout-free streaks in modern baseball?

- Can we characterize the hitters who have long strikeout-free streaks?

To address these questions, we explore data that contains the outcomes of every plate appearances for all batters in the 2016 baseball season. Since it makes sense to focus on everyday position players (excluding pitchers), we focus on the 258 players who had at least 300 plate appearances.

For each player, we find the longest streak of consecutive PAs without a strikeout and Figure 9.7 displays a histogram of these longest "strikeout-free" streaks. Many of these players have longest streaks of lengths 10 to 30, but we do see a few unusually long lengths. The figure shows a vertical line at the value of 50 and there are six players–Mookie Betts, Jose Iglesias, Martin Prado, Nori Aoki, Adam Eaton, and Angel Pagan–who all had longest streaks exceeding 50. Betts actually had a streak of 78 PAs without a strikeout and he continued this streak into the 2017 season.

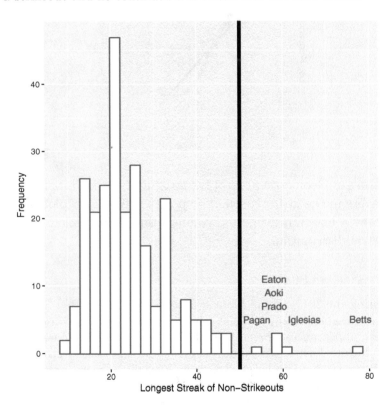

Figure 9.7 Histogram of lengths of the longest strikeout-free streaks for all 2016 players with at least 300 plate appearances. The vertical line at 50 is used to identify six unusually large streak lengths.

One can better understand the significance of these non-strikeout streaks by also looking at the number of plate appearances and strikeout rates for these players. Figure 9.8 displays a scatterplot of the number of PAs and the longest strikeout-free streak for the players with at least 300 PA. The six players with the extreme values are labeled. Looking at the pattern of the smoothing curve, we see that players with more PAs tend to have longer strikeout-free streaks. This makes sense – if a player gets more opportunities to bat, then there will be more opportunities for long streaks. The second interesting observation is that our six players got these long streaks with different numbers of opportunities. Mookie Betts got his streak of 78 in a season where he had 730 plate appearances. In contrast, Nori Aoki had a streak of 58 with a season of only 466 plate appearances. Perhaps the streaky accomplishment of Betts was not that much better than the accomplishment of Aoki since he had 264 more opportunities to hit in this season.

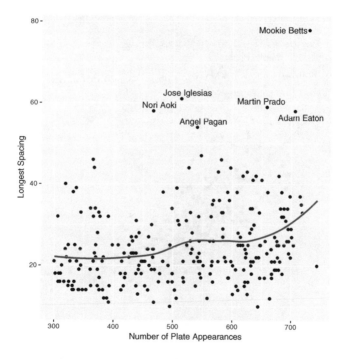

Figure 9.8 Scatterplot of number of plate appearances and longest strikeout-free streak for all 2016 players with at least 300 plate appearances. The smoothing curve indicates that players with more opportunities to hit tend to have longer strikeout-free streaks.

Another way to view these long streaks is to look at the relationship of the longest streak with the strikeout rate defined by the fraction of PAs where there was a strikeout. Figure 9.9 shows a scatterplot of strikeout rates and longest strikeout-free streaks for our 258 players. We see a strong relationship between strikeout rate and longest streak – players who don't strike out much tend to have longer strikeout-free streaks. Mookie Betts' strikeout rate was only 0.11 and so perhaps it is not that surprising that he had a streak of 78. From this perspective, Eaton's streak of 58 strikeout-free PA may be more remarkable in that his overall strikeout rate was 0.16, significantly higher than Betts' strikeout rate. The main point here is that lengths of streaks should be viewed in the context of the number of opportunities and the overall rate of a success.

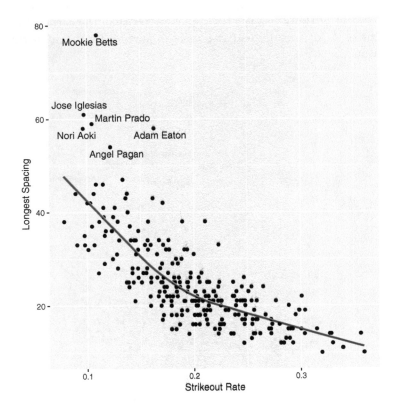

Figure 9.9 Scatterplot of strikeout rates and longest strikeout-free streak for all 2016 players with at least 300 plate appearances. The smoothing curve indicates that players with smaller strikeout rates tend to have longer strikeout-free streaks.

CLUTCH PLAY

In baseball, players are recognized for performing well in particular situations. Fans are familiar with these clutch performances: a batter may hit a home run when the game is tied in the 9th inning, a fielder might make a great catch to preserve his team's one-run lead, and a relief pitcher might enter a game with no outs and the bases loaded and not allow any runs. Particular players such as Reggie Jackson and David Ortiz are famous for their clutch performances. In fact, Jackson had the nickname "Mr. October" since he performed so well during the baseball playoffs.

In this chapter, several methods for measuring a player's streaky performance are described. Is there a method for measuring a player's clutch hitting performance during a season? Here we describe a method for assessing a player's situational hitting – this method will measure how well a player takes advantage of different runners on base and outs situations during a game. Once this measure is described, then we will use it to find the player in the 2016 season who displayed the most "clutch" hitting.

Let's focus on Stephen Piscotty, an outfielder for the St. Louis Cardinals for the 2016 season. In this season, Piscotty had 649 plate appearances including 99 singles, 35 doubles, 3 triples, 22 home runs, 51 walks, and 22 hit-by-pitches. All of these on-base events create runs for his team. But we learned from Chapter 3 that the runs value of, say a single, depends on the runners on base and number of outs. A single with the bases loaded and two outs is certainly more valuable from a runs perspective than a single with the bases empty and one out.

Leverage provides a way of measuring the importance of a runners/outs situation. For a given runners on base and number of outs, one considers all of the possible batting events (out, walk, single, double, triple, home run) and the runs values of all of these possible events. Leverage measures the variability of these run values, taking into account the likelihoods of the different events.

Figure 9.10 displays a scatterplot of the leverages of all possible runner and outs situations against the corresponding situation percentages. We see that the no runners on base situations (labeled "000 2", "000 1", and "000 2") occur frequently, accounting for a total of 57% of the PAs, but they all have small leverage values. The batting event has little impact on the runs value in these situations. In contrast, the runners on base situations have higher leverage values, but these situations occur infrequently. For example, the bases loaded/two outs situation (labeled "123 2" in the figure) has the highest leverage but this situation only occurs about 1% of the plate appearances.

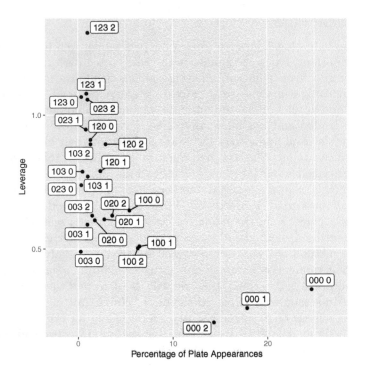

Figure 9.10 Scatterplot of leverages and percentage of total plate appearances of the different runners-on-base and outs situations. This indicates that the high leverage situations don't happen very often in baseball.

During Stephen Piscotty's 2016 season, he had a total of 99 + 35 + 3 + 22 + 51 + 22 = 232 on-base events. If Piscotty is a good situational hitter, then he would plan his on-base events to occur during the situations of high leverage. In other words, he would be more likely to have on-base events during the runners-on-base high leverage, and less likely during the bases empty/low leverage situations.

If there is no relationship between one's hitting and the runners/outs situation, then all possible orderings of the on-base events and outs across these situations are equally likely. This suggests a method of measuring situational hitting. One randomly mixes up the order of the on-base and out events and (assuming standard runner advancement rules) computes the total runs value of these events. This simulation is repeated a large number of times and one computes the mean number of runs produced if one's batting events are not related to the situations. One then computes the difference

$$CLUTCH = Actual\ Runs\ Produced - Mean\ Runs.$$

If the value of $CLUTCH$ is positive, this means that the batter took advantage of the situations – he produced more runs than one would expect if the batting events were independent of the situation.

For all of the players in the 2016 season with at least 500 plate appearances, Figure 9.11 shows a scatterplot of the mean runs produced against the $CLUTCH$ statistic. A horizontal line is placed at the value 0 – points above this line correspond to hitters who displayed some situational hitting. Stephen Piscotty and Adam Duvall (labeled in the figure) have the highest clutch values. Piscotty evidently had many on-base events during high-leverage situations and produced about 11 more runs than one would expect if hitting and the situation were not related.

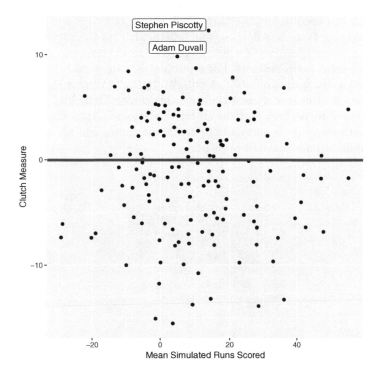

Figure 9.11 Scatterplot of mean runs produced and clutch statistic for all players in the 2016 season with at least 500 plate appearances. Two players are labeled who appeared to hit well during high-leverage situations.

WRAP-UP

Patterns of streaks in hit/out data can be difficult to detect and this chapter has introduced several helpful visualizations. Moving average plots and spacing plots are useful in seeing how players perform in short time periods and detecting long ofer values. Since coin-tossing data can also look streaky, the geometric plot can be used to see if the spacings resemble tosses from a coin – this graph can be used to pick up unusual streaky performances. The strikeout-free streak example was used to illustrate several points. First, players who get more at-bats tend to have longer streaks of non-strikeouts and also players with good bat control tend to have longer streaks. Understanding these patterns makes it easier to understand the meaning of long streaks.

Fans and media will always be fascinated with clutch performances in baseball. Although players will have clutch performances, it is less clear if players have a general tendency to allocate their batting outcomes according to the situation. A special graph was used to find the players who produce more runs than one would expect if their batting was independent of the situation. An outstanding question is whether players can consistently perform clutch over many seasons. For example, no one would doubt that David Ortiz had a number of clutch hitting performances during his career. But it is unclear that over his career, Ortiz had a general tendency to perform well in clutch situations. Given the interest in both streakiness and clutch play, these topics will likely remain popular among baseball researchers.

Bibliography

[1] Cleveland, W. (1994), *Elements of Graphing Data*, Hobart Press.

[2] Fast, Mike (2010), "What the heck is PITCHf/x." *The Hardball Times Annual 2010*, 153–158.

[3] Gould, S. J. (2011). *Full House*. Harvard University Press.

[4] James, B. (1994). *The Politics of Glory: How Baseball's Hall of Fame Really Works*. Macmillan.

[5] Lewis, M. (2004). *Moneyball: The Art of Winning an Unfair Game*. WW Norton & Company.

[6] Marchi, M., and Albert, J. (2013), *Analyzing Baseball Data with R*. CRC Press.

[7] MLB Advanced Media, L.P. (2016). *Statcast*. Available fromhttps://baseballsavant.mlb.com..

[8] R Core Team (2017), "R: A language and environment for statistical computing," R Foundation for Statistical Computing, Vienna, Austria, www.R-project.org.

[9] Wickham, H. (2016), *ggplot2: Elegant Graphics for Data Analysis*, Springer, New York.

[10] Wilkinson, L. (2005), *The Grammar of Graphics*, second edition, Springer, New York.

Index

ability of a team, 106
Almora, Albert, 102
Alomar, Roberto, 24
Angel Stadium, 74
Aoki, Nori, 122, 130
Arrieta, Jake, 56
AT & T Park, 68, 74

ball
    definition of, 46, 60, 62
    in play, 64
ballpark effects, 68
    direction of home run, 73
baseball competition model, 104
baseball playoffs
    1968 format, 102
    2017 format, 102
Baseball Reference, 98
baseball, history of, 1
batting average
    30-day, 113
    after each month of season,
        111
    definition of, 10
    disappearance of .400 AVG,
        12
    history of, 10
    over .300, 113
    predicting final, 117
    shrinking, 117
Betts, Mookie, 129, 130
Biaz, Javier, 102
Biggio, Craig, 24
Bogaerts, Xander, 115
Bonds, Barry, 4, 5, 13, 14, 18, 23
    career trajectory of home
        runs, 18

Bradley–Terry model, 104
Brett, George, 11
Brown, Mordecai, 3
Bryant, Kris, 46

career trajectory, 18
    general pattern, 21
    peak performance, 25
Carew, Rod, 11
Carlton, Steve, 27
championship series, 103
changeup, 57
Chapman, Aroldis, 102
Chase Field, 74
Chicago Cubs, 46, 98
Clemens, Roger, 28
Cleveland Indians, 46, 98
clutch hitting
    measure, 135
clutch performance, 133
Cobb, Ty, 18
Cone, David, 29, 30
consistent hitter, 126
contact percentage, 85
    in zone, 85
    outside of zone, 85
count
    definition of, 46
    hitters', 47
    pitchers', 47
Cueto, Johnny, 90
curve ball, 57
Cy Young award, 27–30

Davis, Rajai, 102
dead-ball era, 3, 6, 7, 13
Detroit Tigers, 102

distance
   of home run, 76
division series, 103

Eaton, Adam, 130
ESPN Home Run Tracker, 70
exit velocity
   of batted ball, 77
   of home run, 74
expected runs scored, 34
   in remainder of inning, 35

FanGraphs, 83, 98
fastball
   four-seam, 57
   two-seam, 57
Fenway Park, 74
Fernandez, Jose, 91
flyout, 77
foul, 62
four-seam fastball, 57
Fowler, Dexter, 46, 99
Foxx, Jimmie, 13

gaps in hitting, 123
Gehrig, Lou, 3, 23, 74
generalized additive model, 79
geometric
   distribution, 127
   plot, 127
Giambi, Jason, 4
Gibson, Bob, 3
Glavine, Tom, 28
Gooden, Dwight, 29, 30
Great American Ballpark, 68
Great Bambino, the, 4
Greenberg, Hank, 13
ground ball, 77
Guidry, Ron, 27
Gwynn, Tony, 11

Halladay, Roy, 29
Hamels, Cole, 56
Helton, Todd, 4
Hendricks, Kyle, 90
hit-by-pitch, 62

hitter
   left-handed, 71
   right-handed, 71
hitters' count, 47
hitting slump, 123
hitting streak, 123
home run
   distance traveled, 71, 76
   exit velocity, 74
   history of, 4
   history of leading, 12
   horizontal angle, 70
   launch angle, 75, 76
horizontal angle
   of home run, 70

Iglesias, Jose, 130
in-play event, 62

Jackson, Reggie, 133
James, Bill, 21
Jeter, Derek, 24
Joss, Addie, 3

Kershaw, Clayton, 56, 58, 62
Keuchel, Dallas, 53
Klein, Chuck, 3
Kluber, Corey, 46
Koufax, Sandy, 27

Lackey, John, 90
launch angle
   definition of, 75
   of batted ball, 77
   of home run, 75, 76
leverage
   of game situation, 133
   of plate appearance, 100
   plot, 100
Los Angeles Angels, 51

Maddux, Greg, 28
Mantle, Mickey, 4, 21, 30, 74
Maris, Roger, 4, 13, 74
Marlins Park, 68
Martinez, Pedro, 29

Mathewson, Christy, 3
Matthews, Eddie, 22
McClain, Denny, 3
McGwire, Mark, 5, 13
McHugh, Collin, 90
Mendoza line, 113
Minute Maid Park, 74
Moneyball, 13
moving average
    definition, 123
    plot, 123
Mr. October, 133
Murderer's Row, 3

normal curve, 104, 113

Oakland Alameda Coliseum, 70
ofer statistic, 121
offensive wins above replace-
    ment, 21
on-base percentage
    definition of, 13
    history of, 13
Ortiz, David, 133
Ott, Mel, 3

Pagan, Angel, 130
park factor, 69
peak performance, 25
Pineda, Michael, 90, 91, 94
Piscotty, Stephen, 133
pitch
    speed, 58
    location, 60, 84
    movement, 58
    outcome, 62
pitchers' count, 47
PITCHf/x system, 55
plate appearance
    definition of, 38
    runs value of, 38
plate discipline
    for batter, 83
    for pitcher, 83
PNC Park, 73
popup, 77

Porcello, Rick, 93
Prado, Martin, 130
prediction error, 117
probability
    about team's ability, 108
    of a base hit, 78
    team reaches milestones,
        107
    winning game, 99
Progressive Field, 73
Pujols, Albert, 30, 51

random hitting data, 125
regression model, 92
Rodriguez, Alex, 4
Rogers Centre, 74
runs expectancy, 98
    matrix, 36
    of ball in play, 51
    passing through a count, 49
runs expectancy matrix, 48
runs potential, 36
runs scored, history of, 2
runs value
    of double, 36, 39
    of home run, 43
    of plate appearance, 38, 48
    of single, 39
    of strikeout, 38
    outcomes of plate appear-
        ance, 41
Ruth, Babe, 3, 4, 13, 14, 23, 74
Ryan, Nolan, 10

Safeco Field, 68, 73
Sale, Chris, 56
Santana, Johan, 27
Scherzer, Max, 53, 91
Schmidt, Mike, 22
Schwarber, Kyle, 46
Shaw, Brian, 102
shrinkage
    estimator, 117
similarity scores, 21
Simmons, Al, 3

simulating a baseball season, 105
slider, 57
slugging percentage, 21
Smoltz, John, 28
smoothing curve, 2
Sosa, Sammy, 5
spacings
    definition, 124
    graph of, 125
St. Louis Cardinals, 102
StatCast system, 67
steroids era, 4
stolen base, history of, 7
streak with no strikeouts, 129
strike
    called, 60, 62
    definition of, 46
    first-pitch rate, 90
    swinging, 62
strikeout
    history of, 9
    no-strikeout streak, 129
    rate, 86, 132
Sultan of Swat, 23
swing outcome, 63
swing percentage, 85
    in zone, 85
    outside of zone, 85
swing-and-miss, 62
Syndergaard, Noah, 91

talent of baseball team, 104
Target Field, 74
Thomas, Frank, 22
three true outcomes, 8
Tomlin, Josh, 90
triples, history of, 6
Trout, Mike, 51, 84
Turner Field, 68
two-seam fastball, 57

walk
    history of, 10
    rate, 87

Walker, Neil, 122
WHIP, definition of, 26
Wild Card game, 103
Williams, Ted, 11, 14, 23
Wilson, Hank, 13
win expectancy, 98
    graph, 98
win probability, 99
    added graph, 101
wins above replacement, 53, 91
    offensive, 21
World Series
    2016 season, 46, 99
    MVP, 102
    probability of winning, 108
Wrigley Field, 74

Yankee Stadium, 68, 73
Yastrzemski, Carl, 3
year of the pitcher, 3, 10
Young, Cy, 3, 18
Yount, Robin, 24, 30

Zobrist, Ben, 102
zone percentage, 85
zone, definition of, 60